Historical Aquaculture
in Northern Europe

Historical Aquaculture in Northern Europe

Edited by
Madeleine Bonow
Håkan Olsén
and Ingvar Svanberg

Södertörn University
The Library
SE-141 89 Huddinge

www.sh.se/publications

© The authors

Cover image: Pond Crucian Carp (*Dammruda*) from Mörkö,
illustrated by Wilhelm von Wright and taken from
Skandinaviens fiskar: målade efter lefvande exemplar och ritade på sten
Stockholm: P. A. Norstedt & Söner, 1836–1857

Cover: Jonathan Robson
Graphic Form: Per Lindblom & Jonathan Robson

Stockholm 2016

Research Report 2016:1
ISBN 978–91–87843–62–4

Contents

Preface

This book is being published in order to highlight a little-known aspect of animal husbandry in former times, namely the keeping, storing and cultivation of crucian carp (*Carassius carassius*), carp (*Cyprinus carpio*), tench (*Tinca tinca*) and other cyprinids in man-made ponds. Aquaculture was an innovation that spread rapidly in northern Europe in late medieval times. Cyprinid ponds continued to be of some importance for the local economies in Scandinavia until the nineteenth century, and have also survived to some extent in regions such as Poland and the southern Baltic region. Although some old ponds remain and traces of others can be seen in the landscape, this historical fish production under human care is very little known in this part of Europe. Cultivation of salmonid fish is of more recent date which is not covered in this book. Pond rearing of brook trout (*Salvelinus fontinalis*), rainbow trout (*Oncorhynchus mykiss*) and brown trout (*Salmo trutta*) started late nineteenth century.

Although an increasing interest in the importance of aquaculture in earlier times has been noted especially in the UK, it has otherwise not been the subject of much research. The aim of this book is to remedy this deficiency. It deals with the variety and complexity that characterize aquaculture in the pre-industrial Baltic region (Scandinavia, the Baltic States and Poland) and the British Isles. Six case studies present historical aquaculture with a special emphasis on cyprinids (crucian carp, carp, tench and other species) that were bred in captivity in man-made ponds. The case studies cover various regions of northern and north-western Europe and show similarities but also differences due to cultural, economic and social circumstances. The introductory section consists of two chapters, which provide a general discussion on the importance of and a possible future for cyprinids in aquaculture, and the role of fishponds in pre-modern monastic economies.

The book is a result of the research project "The Story of the Crucian Carp (*Carassius carassius*) in the Baltic Sea region: History and a Possible Future" at

Södertörn University, sponsored by the Baltic Sea Foundation. Field research was made possible by a grant from C. F. Lundström's Foundation. The contributors represent a variety of disciplines such as archaeology, economic history, ethnobiology, garden history, human geography, limnology, osteology, and zoo physiology. This indicates that research on historical aquaculture can be done within a number of disciplines. Some chapters are based on papers presented at the workshop "History of Aquaculture in Northern Europe" held at the Royal Gustavus Adolphus Academy for Swedish Folk Culture in Uppsala, Sweden, from 3–5 May 2012. We would like to thank our sponsors for making this book possible. Special thanks also go to Professor Richard C. Hoffmann (York University, Canada), the Swedish Museum of Natural History in Stockholm and Uppsala Centre for Russian and Eurasian Studies at Uppsala University.

Stockholm and Uppsala, Spring 2016
Ingvar Svanberg, Madeleine Bonow and Håkan Olsén

Introduction

Håkan Olsén and Ingvar Svanberg

Production of fish in ponds has a long tradition in Europe. In his *De re rustica Libri XII* from the early first century, the Roman author Lucius Iunius Moderatus Columella gives a vivid description of fish farming. As the Roman Empire extended into north-western Europe, the practice of aquaculture also spread. However, there is no evidence that the Romans bred carp in their aquaculture facilities. After the decline of the Roman Empire, the use of fishponds appears to have ceased and for several centuries there was very little interest in cultivating fish in this way. In the eleventh century, due to the rise in human population and overfishing of certain wild stocks of sought after fish, populations of for example sturgeon, salmon, whitefish and trout began to decline (Hoffmann 1996; Makowiecki 2008).

Some definitions are necessary to help us in our discussion on aquaculture. We need to make some clear distinctions about types of aquaculture since much confusion arises from writers not differentiating between natural fish populations in natural or artificial ponds, unselective capture for stocking or storage of wild fish, selective stock and grow operations, human management of breeding and species-specific stocking, and artificial feeding or nutrient management. We deal mainly with the last case, although examples of the other kinds will also be given in following chapters. We do not include marine aquaculture, which is a very recent phenomenon in northern Europe, including Scandinavia (Hoffmann 1996; Bonow and Svanberg 2016).

The possibility to grow and breed other fish species in ponds bought about resurgence in constructing fishponds. However, it would be several centuries before aquaculture production of food fish spread northwards (Figure 1). The construction of fishponds began across Europe, and increased rapidly during the twelfth and thirteenth century. At that time, fishponds were constructed on estates belonging to bishops, monasteries and royalty across England (Hoffmann 1995). During the High Middle Ages, pond-breeding of fish for food developed as a new economic activity in central and Western Europe. It was an innovation that spread rapidly in the fourteenth and fifteenth centuries and finally also reached Scandinavia. It was diffused through both monastic and secular aristocratic channels (Bond 1992). The most important species in central Europe was the common carp, *Cyprinus carpio* L., 1758, which was introduced into some parts of central and Western Europe in the twelfth and thirteenth centuries (Hoffmann 2002).

In northern Europe, however, it was difficult to winter and reproduce carp and it was therefore replaced with the crucian carp, *Carassius carassius* (L., 1758), Figure 2, which was bred in ponds and used as food. Other cyprinids (especially bream, *Abramis brama* (L., 1758)), and other fish species (pike, *Esox lucius* L., 1758) were kept as well, but they seldom reproduced in the ponds. These taxa were therefore merely intended for store-ponds. However, the crucian carp was well-suited to the Swedish climate, it grew quickly if properly tended, and it produced a good yield. By contrast, the common carp was hard-pressed to survive the harsh Swedish winter. Northern Europe's climate does not allow carp to reproduce either. The crucian carp, however, can be maintained under anoxic conditions for months. It is therefore a very useful species as a pond fish under the conditions Scandinavia and other parts of northern Europe offer. The crucian carp was actually popular in pond-breeding also in the Baltic region and Poland (Makowiecki 2008). In Scandinavia, we do not have any evidence until late medieval times (Rasmussen 1959). Sources from the British Isles and Scandinavia show that both clerical and lay landowners constructed and owned fishponds (Svanberg et al. 2012). Details about the development of cyprinid aquaculture will be given in the following chapters covering specific geographical areas. A chapter on Germany and adjacent areas was also planned but the author did not manage to finish it in time for this book.

One of earliest manuals for cultivating cyprinids in fishponds was written by the Swedish friar Peter Magni in 1520. He deals in detail with how to keep crucian carp in ponds. However, it was never printed and therefore did not reach any audience. There are a few printed sixteenth-century treatises on freshwater aquaculture from central Europe. The first one, on carp and pike ponds in Moravia and Silesia, probably written in between 1525 and 40 for Anton Fugger, a landlord and owner of several fishponds, was published in Latin in 1547 by Johannes Dubravius (c. 1486–1533), Bishop of Olmütz in Moravia since 1541. It was translated into English and Polish and was read by noble landowners all over central and northern Europe (Colin 2011, Svanberg and Cios 2014).

The second manual was published in 1573 (second edition in 1605) by Olbrycht Strumieński (?–1609), a nobleman and manager of fishponds in the vicinity of Kraków. He wrote the book at the request of his friends, who valued his experience in aquaculture very highly. In the introduction he stated that »some respectable persons« had written works in Latin, but he does not know them. Due to the popularity of carp aquaculture, his book was plagiarized by Stanisław Stroynowski and published twice, in 1609 and 1636. These manuals give us insight into how aquaculture was conducted in early modern times. Further manuals on the construction and management of fishponds were published in northern Europe in the seventeenth century but these will not be discussed here (Svanberg and Cios 2014).

Domestication of cyprinids seems to have taken place early both in Europe and in China. It is a widespread misunderstanding that the carp, *Cyprinus carpio* L., was domesticated in Asia. It actually occurred in south-east Europe, in the Danube basin, during medieval times. Various domesticated varieties have been available for a long time, especially so-called leather carp and mirror carp (Balon 2004). Domestication of the goldfish (*Carassius auratus*) began in China, probably already during the Song dynasty (Chen 1956). The tench (*Tinca tinca*) has been kept for food production, probably already in late medieval times and it also had a reputation of being good for other cyprinids in the ponds, keeping them healthy. We have no reliable information about when the golden variety became common among the breeding fish (Balon 2004).

Ornamental fishponds

Cyprindids have also been kept in ponds for ornamental purposes (Figure 3). The topic is interesting and a few details can be given here. Although keeping ornamental fish in garden ponds has a long tradition in Europe, the subject has not been extensively researched. Surprisingly little is known about their history as ornamental fish within garden pond culture. As far as we can see, only the goldfish has been discussed at length by historians, although details are still lacking.

Crucian carp (Figure 4) seem to have been common in ornamental ponds in Scandinavia already in the seventeenth century (Figures 5–7). However, with increasing contact with park culture on the continent, various golden varieties of cyprinids were primarily kept for ornamental purposes in garden culture.

Other cultural varieties of indigenous cyprinids have also been kept for ornamental purposes. In addition to the crucian carp, four species of cyprinids dominated in the park culture in older times: carp and its varieties, goldfish, and its varieties, and the golden variety of tench known as golden tench, and the golden variety of ide or orfe, *Leuciscus idus* (L.).

Carp is a stunning pond fish, but due to its sensitivity to cold water has been rarely kept in Scandinavia. The domesticated mirror carp has been kept in larger garden ponds. Colour varieties seem to have been developed in Japan in the early nineteenth century. In 1914, colour breed carps were put on show at an exhibition in Tokyo, which was the beginning of interest in the so-called Koi carp in Japan. A real craze for the beautiful fishes developed. Soon exports of Koi carp began and eventually spread worldwide. The name Koi carp is actually a tautology (Koi meaning 'carp' in Japanese). In Japanese they are referred to as *nishikigoi*, literally meaning 'brocaded carp'.

The goldfish exists in a variety of breeds that can be categorized into several groups according to their body shape. The oldest varieties are the fantail, telescope eye (*demekin*) and *shubunki*. However, many other breeds exist and several of the more extreme ones do not live in ponds but only in aquarium tanks. Goldfish have been kept both in Chinese-style rather large porcelain jars, usually imported from China, later in glass bowls and aquarium tanks, but also in garden ponds. It breeds freely in ponds. According to some reports, the goldfish reached Europe with Portuguese ships from southeast Asia in 1611. Its beauty and colour attracted the attention of royal courts and academies (Hervey and Hems 1948). No native species could remotely compare with the golden newcomer from China. Its presence was recorded in England in 1691 and it spawned for the first time in the Netherlands in 1728 (Tyrbjerg 2006). Few details about its introductions are known though, and therefore some details about its early presence in Sweden may be of interest, based mainly on research by Svanberg (2007).

In Sweden, the goldfish has been known since the 1740s. With the help of Swedish ambassador Nils Palmstierna, the Swedish Academy of Sciences received a preserved goldfish, which was desiccated by Carl Linnaeus and described in the proceedings from the academy in 1740. Interestingly enough, Linnaeus also published some information about its care. It was still a fish kept in bowls, not in ponds, in Scandinavia, and it should be given fresh water two-three times a week. It should be fed with biscuit, egg yolks, dried lean pork and small snails. It is however unclear if Linnaeus had any personal experience of live goldfish (which he actually might have had from his years in the Netherlands). However, the information he gives about the care of goldfish bears witness that he had received information from someone who knew. One of the depicted gold fish was obviously a fantail. Linnaeus never ceased to be fascinated by the beauty of the goldfish. In 1744, the merchant Carl Gyllenborg donated "Chinese goldfish" to Linnaeus, but they were preserved in ethanol. Linnaeus was eager to get hold of live fish. In November 1745, he wrote instructions for his pupil Christopher Tärnström that he should bring back goldfish for the queen from his tour to East India. However, Ternström unfortunately drowned in south-east Asia and was never able to complete the task. Linnaeus's enthusiasm for the goldfish was undiminished. He received further preserved goldfish for his zoological collections from the royal court and King Adolph Fredric and his queen probably kept goldfish in the ponds at the royal palace of Drottningholm.

In 1759, an opportunity arose for Linnaeus to receive a goldfish. One of his pupils, the physician Pehr Bjerchén, was staying in London and kept up a lively correspondence with Linnaeus. He asked Bjerchén to find goldfish for him. Linnaeus became very enthusiastic when he received a letter telling him that Bjerchén had traced down a breeder in his colleague Richard Guy. He had a pond at his country estate with 50 to 60 goldfish. Guy of course was very careful about his rare fish, but Bjerchén persuade him to donate a couple to the world-famous professor in Uppsala. Bjerchén brought them back by ship to Gothenburg. Linnaeus wrote impatiently in September 1759, "Do please send the goldfish already tomorrow with a vessel to Uppsala; they do not freeze that easily; so I will be able to once again see them, something I have dreamed of all my days but never hoped. Let the skipper ask what price he wants, only I get them alive". He finished the letter, "God give I had them alive in my orangery". Apparently, the bespoke thrifty Linnaeus was prepared to pay quite a sum to get his coveted goldfish. Bjerchén sent the goldfish together with detailed information about how to take care of them. It is a very interesting sheet of instructions with information about keeping the bowl clean, the importance of fresh water and how to feed the goldfish. In October 1759, one fish was already dead, but the other specimen survived at least the winter and was alive the next year. Dr Guy in England promised to provide Linnaeus with further specimens. More goldfish were offered by Job Baster, a Dutchmen who was breeding them in open-air ponds. It is, however, unclear if Linnaeus received any more live goldfish (Svanberg 2007).

In the late eighteenth century, the goldfish was widely kept in south-western Europe. In 1780, Louis Edme Billardon de Sauvigny published a famous book about the Chinese goldfish with nice illustrations (Sauvigny

1780). In his book on companion animals, Johannes Matthäus Bechtein (1795: 222–226) gives a rather detailed description of keeping goldfish.

The golden tench is mentioned by Marcus Bloch (1777) and seems to have originated in central parts of Europe, probably Bohemia and Silistria. It spread early to other parts of Europe and was kept as an ornamental fish for garden ponds and pleasure waters (Shaw and Stephens 1804: 217–218). Another domesticated fish is the golden ide or orfe, which probably comes from Germany and seems to have been common in continental fishponds already in the seventeenth century. Leonard Baldner named it *Goldgelbe Rottel* in his manuscript *Recht natürliche Beschreibung und Abmahlung der Wasservögel, Fische, vierfüssige Thieren, Insecten, und Gewürmb, so bey Strassburg in den Wässern seynt, die ich selber geschossen und die Fisch gefangen* from 1666 (Lauterborn 1901).

The increasing availability of goldfish since the 1960s through the flourishing pet trade and the relatively low prices of large aquarium tanks to keep them during the winter have been important in this process. Still thousands of garden ponds are constructed in Scandinavian small gardens and this has increased the demand for suitable fish species. The expensive but long-lived koi carp have become more and more popular and they can also winter in the ponds if they are deep enough. Many different species of fish are available for garden ponds in northern Europe today through the pet trade. In addition to various goldfish varieties and so-called koi carp, species like sterlet, *Acipenser ruthenus* (L.), grass carp, *Ctenopharyngodon idella* (Valenciennes) and the small gudgeon, *Gobio gobio* (L.) can also be found.

The future of cyprinid culture

The depletion of marine fish stocks due to overfishing and the ongoing changes in climate will require changes in fisheries and aquaculture. One of several necessary changes in aquaculture will be to increase the use of omnivorous species that have a high degree of plant materials in their diet. The amounts of marine proteins and oils in fish feed have to decrease and be replaced by other sources. Aquaculture based on predatory fish, such as salmonid fishes, are not sustainable, as fish are fed by fish proteins (Naylor et al. 2000). According to FAO (2006), 53 per cent of the global fishmeal and 87 per cent of the fish oil was consumed by salmonids, marine fish and shrimp in aquaculture, and fish meal and oil production has stabilized and is not increasing due to depleting resources ("the fish meal trap"). A recent FAO report deals with the increasing problems with feeding fish with fish in aquaculture (FAO 2011).

Scientific research is going on to develop commercial vegetarian feed to carnivorous species such as salmon (Powell 2003), but that is like "crossing the stream to get a bucket of water", as there are suitable omnivore fish species such as cyprinids to begin with and they do not contribute to the depletion of marine fish stocks (Naylor et al. 2000). World-leading fishery scientists have emphasised the seriousness of the situation and point out that it is important to increase production of species with a more vegetarian diet. Plant materials added for instance to salmon feed have resulted in various problems such as intestinal inflammation (e.g. Urán et al. 2008) and reduced palatability resulting in lower growth rate (e.g. Pratoomyot et al. 2010). The aquaculture industry might later have a salmon feed that contains a majority of plant materials but the development process to get a safe feed takes time. There are some promising results published (De Santis et al. 2016). It is also important not to use crops such as soybean to feed fish when they can be used as food for humans. Aquaculture based on cyprinids, on the other hand, should be more sustainable from the beginning as their diet is more varied and not based on fish. Their lower demand for high quality water regarding oxygen content and turbidity should also be in their favour compared to salmonids. Cyprinid fish were popular as food in Sweden before World War II and they are still popular in other parts of Europe. The demand for fish food in China and other "developing" countries with growing economies will increase significantly with an intensifying race for marine fish resources. There will be more people who want to share the resources (Pinnegar and Engelhard 2008).

The increase in temperature due to the global climate change will have detrimental effects on cold- and cool-water species fish in the northern hemisphere (Schiermeier 2004; Ficke et al. 2007). In addition to the effects of the water temperature on the limnology, for example, stratification and oxygen content depletion due to eutrophication and bacterial metabolism, the physiology and survival of the fish is affected in species-specific ways. Cold-water fish such as the economically important salmonids will probably suffer from the increase in water temperature as they have an optimal in the low temperature range for reproduction, growth and activity (Jonsson & Jonsson 2009). Even if food availability increases to cover the higher metabolic rate, it will not be sufficient for an increase in growth rate as there will be no corresponding increase in foraging activity (Brett 1971). A recent study has shown that Atlantic salmon (*Salmo salar*) lose their appetite when there is an increase in water temperature from 14 to 18–19 °C (Hevrøy et al. 2012). At the same time as their appetite decreased, blood plasma levels of the appetite-stimulating hormone ghrelin also fell. In addition to slow growth rate, fish also had problems to absorb lipids from the feed and instead used their own energy resources by depleting their fat resources.

Stenothermal cold water species will also be negatively affected by increased eutrophication, as the oxygen content in the hypolimnion used during summer as a refuge will decrease and be replaced by warm water species tolerating low oxygen levels. In aquaculture, salmonid fish will be replaced by warm water preferring and hypoxia-tolerant fish. In our opinion, it is important to begin planning for such a change already now. Due for instance to some of the cyprinid species' high temperature resistance, for example, crucian carp (Horoszewicz 1973), it will be one of the most suitable species in the Baltic area. Different species of Chinese carp, such as grass carp are widely distributed all over the world (Lin and Peter 1991) and give rise to different problems after escaping into pristine environments. The problems in the USA, Australia and New Zealand with common carp as a serious pest in certain natural waters (Stuart et al. 2006) may make it less suitable compared to the native species when the temperature is increasing. Common carp are also carriers of various viruses and there is a risk of diseases being introduced into pristine environments (Gozlan et al. 2005; Hellström et al. 2012). To decrease the risk of alien species affecting the environments, local species should be adapted to aquaculture. Evaluating new species to be used in aquaculture is also stressed by FAO (2010). The development of production systems with crucian carp with closed but low energy systems, like bio-floc ponds (Kiessling 2009; Wang et al. 2015), are therefore a prerequisite for increased production in the Baltic area. Furthermore, the crucian carp with its potential to efficiently feed on bio-floc based on organic waste streams, i.e. feed, has the potential be a net food producer, transforming organic waste into high quality human food (Kiessling 2009, 2011). In fact, crucian carp are used in polyculture systems in China together with various carp species (Lin and Peter 2001). The probiotic capacity of the bio-floc (Kiessling 2009), together with no external water exchange also offers a farming system with low risk of internal disease problems and infecting wild fish in surrounding waters.

Fast-growing strains of crucian carp could have a good potential in aqua-culture as a single species or together with other cyprinids. From the six-teenth to the eighteenth centuries, crucian carp were an important food source in Scandinavia and the Baltic states (Bonow and Svanberg 2011). The fish were kept in ponds and due to their anaerobic respiration they could survive anaerobic winter conditions when other species died (Shoubridge and Hochachka 1980; Piironen and Holopainen 1986). Keeping crucians was important for food production and was used during periods of abstinence, but crucian carp farming continued after the Reformation. During his journey to Skåne 1749, Carl Linnaeus was amazed by the extensive system of ponds used for farming crucian and common carp. Crucian carp was appreciated as food in the households of the aristocracy and the upper classes (Svanberg 2006). Crucian carp and the closely related gibel carp, *Carassius gibelio* (Bloch) are considered to be two of the most important aquaculture species in the world (Billard and Berni 2004). Ranked globally, production is in sixth place among fresh water fish with a production of 1.7 million tons in 2002 with a commercial value of 1.2 billion US dollars (FAO, Fisheries and Agricultural Department; www.fao.org/). By 2006, the production of *Carassius spp.* (stated as *Carassius carassius*) had increased to 2.1 million tons and advanced to fifth place (Subasinghe 2009). In Europe, there is some production of crucian carp in Estonia, Latvia, Belarus, Slovenia and Moldova. There is also increasing interest in Poland to cultivate the species (Rybcsyk and Szypula 2005). Armenia, Azerbaijan and Taiwan are other major producers. Most crucian and gibel carp, however, produced by fish farms in China (99.6 % of the global total).

Crucian carp is an omnivore species that feeds on organic detritus, filamentous algae, small benthic animals and pieces of aquatic weeds. They have an excellent taste and high meat quality, but it has large numbers of fine inter-muscular bones. These problems should be less when the fish are bigger, 500 g. Individuals of 4–5 kg have been caught in Finland and eastern parts of Europe. There are great differences in growth rate in crucian carp between different locations due to environmental conditions and population densities, and probably also due to genetic differences (Szczerbowski and Szczerbowski 2002). The species has a strong disease and parasite resistance compared to other species, both wild (Karvonen et al. 2005), and cultured and this entails that antibiotics and other chemicals can be used at minimum levels, which is important as there are increasing risks and concerns with resistant bacteria in the environment (Medical Products Agency, *Läkemedelsverket* 2004).The species has the prerequisites to be an "organic species" and by using stocks with genetic variability and selective breeding it will be possible to develop a stock with improved growth rate. Salmonid aquaculture continues to be surrounded by disease problems and chemical treatment to control microbial and invertebrate parasites (Rosenberg 2008).

References

Balon, E. K. 2004. About the oldest domesticates among fishes. *Journal of Fish Biology* 65 (Supplement A): 1–27.

Bechstein, J. M. 1797. *Naturgeschichte der Stubenthiere*. Gotha.

Billard, R. and Berni, P. 2004. Trends in cyprinid polyculture. *Cybium* 28: 255–261.

Billardon de Sauvigny, L. E. B. 1780. *Histoire naturelle des dorades de la Chine*. Paris.

Bloch, M. 1785. *Naturgeschichte der ausländischen Fische*. Berlin.

Bond, J. 1992. The fishponds of Eynsham Abbey. *The Eynsham Record* 9: 3–12.

Bonow, M. and Svanberg, I. 2011. »Säj får jag dig bjuda ur sumpen en sprittande ruda«: en bortglömde läckerhet från gångna tiders prästgårdskök, pp. 147–169 in Bonow, M. and Rytkönen, P. (eds.) *Gastronomins (politiska) geografi*. Stockholm.

Bonow, M. and Svanberg, I. 2012. Uppländska ruddammar: ett bidrag till akvakulturens kulturhistoria. *Uppland* 2012: 123–152.

Bonow, M. and Svanberg, I. 2016. Monastiska fiskdammar i senmedeltida Sverige, pp. 262–280 in Bonow, M, Gröntoft, M, Gustafsson, S and Lindberg. M. (eds.) *Biskop Brasks måltider: svensk mat mellan medeltid och renässans*. Stockholm.

Brett, J. R. 1971. Energetic responses of salmon to temperature: a study of some thermal relations in the physiology and freshwater ecology of sockey salmon (*Oncorhynchus nerka*). *American Zoologist* 11: 99–113.

Brönmark, C., and Miner, J. G. 1992. Predator-induced phenotypical change in body morphology in crucian carp. *Science* 258: 1348–1350.

Chen, S. C. 1956. A history of the domestication and the factors of the varietal formation of the common goldfish, *Carassius auratus*. *Scientia Sinica* 5: 287–321.

Colin, N. 2011. *The History of Aquaculture*. Ames, IA.

Columella, L. J. M. 1948. *De re rustica*. London.

De Santis, C., Tocher, D. R., Ruohonen, K., El-Mowafi, Martin, S. A. M., Dehler, C. E., Secombes, C. J. and Crampton, V. 1916. Air-classified faba bean protein concentrate is efficiently utilized as a dietary protein source by post-smolt Atlantic salmon (*Salmo salar*). *Aquaculture* 452: 169–177.

Denverk, D. R., Morris, K., Lynch, M., Vassilieva, L. L. and Thomas, W. K. 2000. High direct estimate of the mutation rate in the mitochondrial genome of *Caenorhabditis elegans*. *Science* 289: 2342–2344.

FAO Fisheries and Aquaculture 2006. *The State of World Fisheries and Aquaculture*. Roma.

FAO Fisheries and Aquaculture 2011. *The State of World Fisheries and Aquaculture*. Roma.

Ficke, A. D., Myrick, C. A. and Hansen, L. J. 2007. Potential impacts of global climate change on freshwater fisheries. *Reviews in Fish Biology and Fisheries* 17: 581–613.

Gozlan, R. E., St-Hilaire, S., Feist, S. W., Martin, P. and Kent, M. L. 2005. Biodiversity: disease threat to European fish. *Nature* 435: 1046.

Gross, R., Kohlmann, K., Kersten, P. and Murakaeva, A. 2005. Phylogenetic relationships of wild and farmed common carp (*Cyprinus carpio*) stocks based on the mitochondrial DNA polymorphisms: implications for taxonomy and conservation. *Aquaculture* 247: 16–17.

Guo, X. H., Liu, S. J., and Liu, Y. 2007. Evidence for maternal inheritance of mitochondrial DNA in allotetraploid. *DNA Sequence* 18: 247–256.

Hanfling, B., Bolton, P., Harley, M., and Carvalho, G. R. 2005. A molecular approach to detect hybridisation between crucian carp (*Carassius carassius*) and non-indigenous carp species (*Carassius spp.* and *Cyprinus carpio*). *Freshwater Biology* 50: 403–417.

Hartl, D. L. and Clark, A. G. 1997. *Principals of Population Genetics*. Sunderland, MA.

Hellström, A. 2010. Sjukdomar som hotar svensk fisk. *SVA VET Tema: Sjö och hav.* No. 1: 6–7.

Hervey, G. F. and Hems, J. 1948. *The Goldfish*. London.

Hevrøy, E. M., Waagbø, R., Torstensen, B. E., Takle, H., Stughaug, I., Jørgensen, S. M., Torgersen, T., Tvenning, L., Susort, S., Breck, O. and Hansen, T. 2012. Ghrelin is involved in voluntary anorexia in Atlantic salmon raised at elevated sea temperatures. *General Comparative Endocrinology* 175: 118–134.

Hoffmann, R. C. 1995. Environmental change and the culture of common carp in medieval Europe. *Guelph Ichtyology Review* 3: 57–85.

Hoffmann, R. C., 2002. Carp, cods, connections: new fisheries in the Medieval European economy and environment, pp. 3–55 in Henniger-Voss, M. J. (ed) *Animals in Human Histories: the Mirror of Nature and Culture*. Rochester NY.

Holopainen, I. J., Tonn, W. M., and Paszkowski, C. A. 1997. Tales of two fish: the dichotomous biology of crucian carp (*Carassius carassius* (L.)) in northern Europe. *Annales Zoologici Fennici* 34:1–22.

Horoszewicz, L. 1973. Lethal and disturbing temperature in some fish species from lakes with normal and artificially elevated temperature. *Journal of Fish Biology* 5: 165–181.

Jonsson, B. and Jonsson, N. 2009. A review of the likely effects of climate change on anadromous Atlantic salmon *Salmo salar* and brown trout *Salmo trutta*, with particular reference to water temperature and flow. *Journal of Fish Biology* 75: 2381–2447.

Karvonen, A., Bagge, A. M. and Valtonen, E. T. 2005. Parasite assemblages of crucian carp (*Carassius carassius*) – is depauperate composition explained by lack of parasites exchange, extreme environmental conditions or host unsuitability? *Parasitology* 131: 273–278

Kiessling, A. 2009. Feed – the key to sustainable fish farming, pp. 303–322 in Ackefors, H., Cullberg, M. and Wramner, P. (eds.). *Fisheries, Sustainability and Development*. Stockholm.

Kiessling, A. 2011. Mikrober för fiskfoder. *Miljöforskning. Formas tidning för ett hållbart samhälle* April 2011. http://miljoforskning.formas.se/sv/Nummer/April-2011

Lauterborn, R. 1901. Das Vogel-, Fisch- und Thier-Buch des Strassburger Fischers Leonhard Baldner. *Naturwissenschaftliche Wochenschrift* 16: 432–437.

Lin, H. R. and Peter, R. E. 1991. Aquaculture, pp. 590–622 in Winfield, I. J. and Nelson, J. S. (eds.). *Cyprinid Fishes: Systematics, Biology and Exploitation.* London, New York, Tokyo, Melbourne and Madras.

Lin, X. and Peter, R. E. 2001. Somatostatin and their receptors in fish. *Comparative Biochemistry and Physiology* 129B: 543–550.

Makowiecki, D. 2008. Exploitation of early Medieval aquatic environments in Poland and other Baltic Sea countries: an archaeozoological consideration, pp. 753–777 in *L'Acqua Nei Secoli Altomedievali. Spoleto, 12–17 Aprile 2007.* Spoleto.

Medical Product Agency 2004. Miljöpåverkan från läkemedel samt kosmetiska och hygeniska produkter. *Rapport från Läkemedelsverket.* pp 1-169.

Naylor, R. L., Goldburg, R. J., Primavera, J. H., et al. 2000. Effect of aquaculture on world fish supplies. *Nature* 405, 1017–1024.

Piironen, J. and Holopainen, I. J. 1986. A note on seasonality in anoxia tolerance of crucian carp (*Carassius carassius* (L.)) in the laboratory. *Annales Zoologici Fennici* 23: 335–338.

Pinnegar, J. K. and Engelhard, G. H. 2008. The "shifting baseline" phenomenon: a global perspective. *Reviews in Fish Biology and Fisheries* 18: 1–16.

Powell, M. D. 2003. Eat your veg. *Nature* 426: 378–379.

Pratoomyot, J., Bendiksen, E. Å., Bell, J. G. and Tocher, D. R. 2010. Effects of increasing replacement of dietary fishmeal with plant protein sources on growth performance and body lipid composition of Atlantic salmon (*Salmo salar* L.). *Aquaculture* 305: 124–132.

Rasmussen, H. 1959. Fiskedamme o. Fiskeopdræt, pp. 307–309 in *Kulturhistoriskt lexikon för nordisk medeltid* vol. 4. Malmö.

Rosenberg, A. A. 2008. The price of lice. *Nature* 451: 23–24.

Rybcsyk, A. and Szypula, J. 2005. Age as well as length and weight growth of crucian carp from the Szczecin lagoon and the Leszczynskie Lakeland. *Electronic Journal Agricultural Universities* 8: 1–10.

Schiermeier, Q. 2004. Climate findings let fishermen off the hook. *Nature* 428: 4.

Shaw, G. and Stephens, J. F. 1804. *General Zoology, Or Systematic Natural History,* Volume 5:1. London.

Shoubridge, E. A. and Hochochka, P. W. 1980. Ethanol: Novel end product of vertebrate anaerobic metabolism. *Science* 209: 308–309.

Stuart, I. G., Williams, A., McKenzie, J. and Hold, T. 2006. Managing a migratory pest species: a selective trap for common carp. *North American Journal of Fisheries Management* 26: 888–893.

Subasinghe, R. 2009. Aquaculture development: the blue revolution, pp. 281–302 Ackefors, H., Cullberg, M. and Wramner, P. (eds.). *Fisheries, Sustainability and Development.* Stockholm.

Svanberg, I. 2007. »Deras mistande rör mig så hierteligen«: Linné och hans sällskaps-djur. *Svenska Linnésällskapets Årsskrift* 2007: 11–108.

Svanberg, I., Bonow, M. and Olsén, H. 2012. Fish ponds in Scania, and Linnaeus's attempt to promote aquaculture in Sweden. *Svenska Linnésällskapets Årsskrift* 2012, pp. 83–98.

Szczerbowski, J. A. and Szczerbowski, A. J. 2002. *Carassius carassius*, pp. 43–78 in Banarescu, P. M. and Paepke, H-J. (eds.). *The Freshwater Fishes of Europe*, vol. 5/II: Cyprinidae 2:2. Wiesbaden.

Tybjerg, H. 2006. Guldfiskens tidlige historie i Norden. *Svenska Linnésällskapets Års-skrift* 2006, pp. 134–153.

Urán, P. A., Goncalves, A. A., Taverne-Thiele, J. J., Scharma, J. W., Verreth, J. A .J. and Rombout, J. H. W. M. 2008. Soybean meal induces intestinal inflammation in common carp (*Cyprinus carpio* L.). *Fish & Shellfish Immunology* 25: 751–776.

Wang, G., Yu, E., Xie, J., Yu, D., Li, Z., Luo, W., Qui, L. and Zheng, Z. 2015. Effect of C/N ratio on water quality in zero-water exchange tanks and the bifloc sup-plementation if feed on the growth performance of crucian carp, *Carassius aura-tus*. *Aquaculture* 443: 98–104.

Figure 1: Capturing fish in a pond. Engraving from 1582 by Hans Bol. (Rijksmuseum, Amsterdam)

Figure 2: One of several big crucian carp caught in Östhammar on the Swedish Baltic coast in June 2012 (range 507–1820 g). The specimen was 1260 g with a total body length of 368 mm. The fish was prepared according to an old recipe and had a mild pleasant flavour. (Photo: Håkan Olsén)

Figure 3: A crucian carp pond at the author Selma Lagerlöf's estate Mårbacka in Värmland. (Photo: Ingvar Svanberg, 2014)

Figure 4: A crucian carp caught in a pond at Skabersjö castle in Scania. The specimen was 95 g with a total body length of 179 mm. (Photo: Håkan Olsén)

Figure 5: View from one of the ponds at Skabersjö castle with a barn in the background. The pond is the one located in the lower part of the maps in Figure 6 and 7. (Photo: Håkan Olsén)

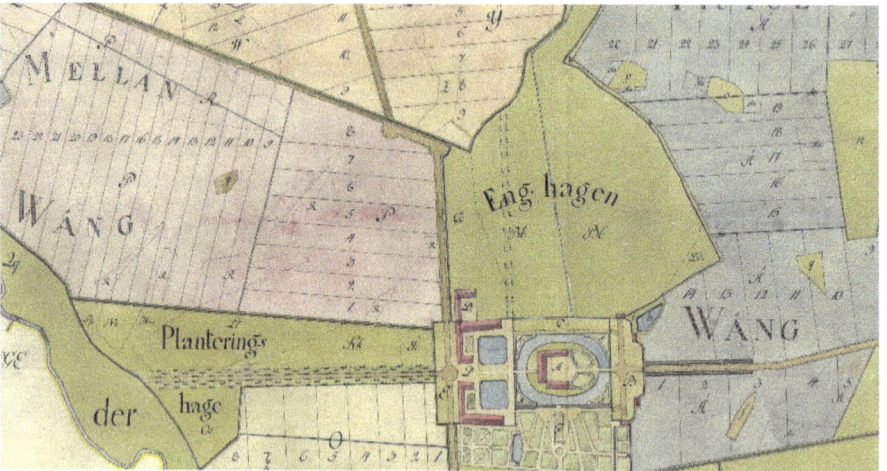

Figure 6: Skabersjö castle in 1758. The castle is surrounded by a moat. In front of the castle there are two squared ponds. Behind the ponds are two barns, in red. The pond located in lower part of the map is shown in Figure 5.

Figure 7: Google map over Skaberjö castle in year 2011. The moat, ponds and barns from 1758 are still present.

CHAPTER 1

Fishponds in the Monastic
Economy in England

James Bond

In any historical investigation, there are good times and bad times to attempt a synthesis. My first attempt to summarise our understanding of monastic fisheries in England and Wales was published at an unlucky moment, when challenges to inherited views were just emerging (Bond 1988).

Until that time it had seemed a reasonable supposition that monks, because of their special dietary requirements, had led the way in introducing fishponds into Britain. It was also widely believed that monastic fishponds supplied food requirements for fast days and the season of Lent, especially in inland areas where sea fish were supposedly less readily obtainable. A third assumption was that the extent and complexity of many monastic fishponds reflected a level of investment which was surely excessive for internal subsistence needs alone, so they must have been designed for breeding and storing fish for commercial sale. None of those beliefs has survived critical reassessment.

The balance of evidence now indicates that fishponds were introduced into Britain after the Norman Conquest (1066) as a secular aristocratic initiative rather than a monastic innovation. Some of the earliest ponds on monastic holdings were actually donated by lay benefactors, and there is no indication that English monasteries adopted a policy of extensive pond construction before the later twelfth century. Investigations of monastic accounts and excavations of kitchen middens have now demonstrated beyond question that marine species were extensively marketed, even deep into the English midlands, and contributed far more to the regular monastic diet than freshwater fish (Bond 2001: 72–74; 2004: 183–210). Despite the efforts which went into their construction, the output from monastic ponds can rarely have been sufficient to meet even basic subsistence needs, let alone producing regular surpluses for commercial sale. Other explanations must, therefore, be sought.

Fish in the monastic diet

The Rule of St Benedict, the most widely-applied monastic rule throughout north-western Europe, did not literally promote the consumption of fish: it merely required abstinence from the flesh of four-footed animals by all but the sick. In England, despite its endorsement by the *Regularis Concordia* and by Archbishop Lanfranc's statutes, strict observance was not universal before the twelfth century. When Simeon, first Norman prior of Winchester (d. 1082), found the monks there eating meat, he persuaded them to change their ways by having the cook prepare exquisite dishes of fish (Knowles 1963: 459–460). Thereafter, observance of customary fish-days every Wednesday, Friday and Saturday and throughout the season of Lent became well-established among the English Benedictines, and was generally respected through to the Dissolution, though the prohibition of meat-eating on other days was relaxed later in the middle ages. Different religious orders developed dietary regulations of varying severity. There was once a belief that the early Cistercians abstained from fish entirely, and Hockey (1970: 50) states that fish only became a regular component of their diet in the late thirteenth century; these views can no longer be upheld (McDonnell 1981: 21–23; Currie 1989: 151). Even when strict abstinence from meat was no longer enforced, marine fish remained readily available and cheap; and economic pressures ensured that fish continued to make a major contribution to monastic diet right up to the Dissolution (Bond 2001: 54–55, 72–74). At Winchester Priory fish formed the main course of the main meal on 165 out of 278 days between December 12th 1514 and September 19th 1515, 59 per cent of the documented days (Kitchin 1892).

Both documentary and archaeological evidence make it abundantly clear that sea fish were consumed in far greater quantities than freshwater fish. Notes of purchases of large numbers of herring and cod abound in monastic kitcheners' accounts, and in 1491 over 45 per cent of Winchester Priory's total food expenditure was on marine fish (Kitchin 1892: 309–362). Even on sites far inland, deposits of monastic food refuse have consistently yielded far greater quantities of bones of sea-fish than of freshwater species (Bond 1988: 70, 74–78, 2001: 72–74, 2004: 183–187). In turn, river fisheries provided far more freshwater fish than managed fishponds. Why, then, are fishponds such a prominent feature on many monastic sites?

The answer almost certainly lies in the concept of exclusivity. Fishponds were costly to construct and to maintain. Pike (*Esox lucius* L.) and bream (*Abramis brama* (L.)), their most esteemed products, were considerably more expensive on the open market than herring or cod (Dyer 1988: 30–32). Grants to monastic communities of fishing rights in rivers and secular ponds sometimes imposed conditions that the catch should be reserved for particular feasts and anniversary commemorations of the deaths of members of the patronal family: in the 1090s William de Warenne, Earl of Surrey, allowed the Cluniac monks of Lewes to fish in his waters for the great feasts and for the memorial of his parents (Salzman 1932: 19). Another purpose was the entertainment of important visitors: the large fishpond of Battle Abbey was fished out in 1275 in anticipation of the king's arrival (Searle and Ross 1967: 42). Fishponds were never intended to supply the entire subsistence needs of monastic communities, let alone to produce revenue from sales; their primary purpose was to provide more prestigious meals for special occasions.

In support of this contention, Currie's calculations on productivity are instructive. He argues that yields from monastic ponds remained relatively meagre, primarily because the fish were left to forage on natural resources. Without supplementary feeding, a pond one acre (0.4 ha.) in extent was able to support only about 90 kg of bream; but, since bream take five years to reach edible size, it could produce only about 18 kg per annum. Assuming 175 fish days a year, at which each monk would receive a minimum of 0.17 kg (0.23 kg of unprepared weight) of fish per day, in order to meet its entire needs, a small house of ten monks would need to produce 385.5 kg each year, which would require 8.5 ha. under water; a large house of 40 brethren would require 36.5 ha. (Currie 1988: 275–276, 1989: 154–155). Furthermore, the needs of guests, corrodians and monastic servants also had to be met. Even allowing for catches from outlying estates, few monasteries in Britain had ponds on this scale. Indeed, some previous estimates have proved over-generous: an earlier suggestion that the monks of Byland Abbey may have controlled up to 75 ha. of ponds must now be reduced in the light of recent survey work (Jecock et al. 2011). Despite possessing extensive fishponds, the Augustinian canons of Waltham and the Cistercian monks of Beaulieu were heavily in debt to London fishmongers in the 1340s (Close R., 1343–1345: 229, 474; Currie 1989: 157).

The chronology of monastic fishponds

English monastic chronicles, charters and accounts contain many incidental references to fishponds (Figure 1.1), but they fail to provide a secure guide to the chronology of construction.

The earliest purported references to fishponds on any monastic property occur in two charters claiming to record royal grants of land to the Benedictine abbey of Abingdon in AD 958–9 and AD 968. Both charters have boundary perambulations attached, referring to the same fishpond (*stirigan pole/strygan pol*). This has been identified with an artificial rectangular pond within the modern parish of Appleton-with-Eaton (Crawford 1930). Unfortunately both charters are of dubious authenticity, and the bounds are certainly later than their nominal date. Neither, therefore, proves the existence of artificial monastic fishponds as early as the tenth century (Sawyer 1968: no. 673 and 757).

The Domesday Survey compiled by order of William the Conqueror in 1086 contains only two records of fishponds at Benedictine abbeys: Bury St Edmunds had two *vivaria vel piscinae*, St Albans one *vivarium piscium* (DB, folios 372, 135v). The Domesday record of fisheries is, however, inconsistent, and it is entirely possible that further monastic ponds may have escaped notice. The appearance in Britain of regular canons and various reformed monastic orders during the twelfth century opened up new links with the continent, which perhaps contributed to the wider adoption of pisciculture.

Records of fishponds on monastic properties begin to proliferate only after the middle of the twelfth century. Many of the earliest examples were pre-existing ponds given to monasteries by lay benefactors. A writ of Henry II permitted the monks of Selby Abbey to have a fishpond "which existed when the abbey was founded", i.e. before 1070 (Farrer 1914: 366). William Ferrers, Earl of Derby, gave the monks of Tutbury a fishpond at Stanford-under-Needwood around 1170 (Saltman 1962: 58). Religious houses were also often granted special fishing privileges in ponds belonging to their patrons: Geoffrey de Clinton, founder of the castle and priory at Kenilworth, confirmed to the priory canons before 1173–4 the right to fish "with boats and nets" in the large artificial lake alongside the castle on Thursdays (Dugdale 1730: i, 238b). Currie (1989: 147–148, 167–168) has listed a dozen examples of monastic houses acquiring from secular lords rights in or possession of fishponds before 1200.

The first known record of a monastic initiative in construction is the fishponds made by Abbot Adam (1160–89) on the lands of Evesham Abbey, though, unfortunately, the chronicler does not locate them (Macray 1863: 101). Fishponds are also recorded on the estates of Pershore Abbey and Fountains Abbey before 1200 (Moger 1924: 158; Farrer 1914: 387, 403).

References to monastic fishponds become much more numerous during the thirteenth century. To some extent this reflects the increasing range of available sources; but it also coincides with a period of direct exploitation, when enterprising abbots were investing in a wide range of improvements to their lands. Abbot Randulph (1214–29) undertook many new works on the Evesham Abbey estates, including the making of new fishponds at Evesham itself and on at least five outlying demesnes (Macray 1863: 261). The Cistercians of Waverley completed construction of a fishpond some 5 km south of the abbey in 1250–51 (Luard 1865: 144). The Cistercians of Beaulieu were constructing a causeway alongside their large pond at Sowley in 1269–70 (Hockey 1975: 257, 302). Though the regime of watermills was not ideally suited to raising fish, Stoneleigh Abbey's millpond at Cryfield was producing perch (*Perca fluviatilis* L.), roach (*Rutilus rutilus* L.), bream (*Abramis brama*), tench (*Tinca tinca* (L.)) and pickerel (*Esox lucius* L.) in the 1380s (Hilton 1960: 220–201).

Because watercourses so often formed property boundaries, it was a common occurrence for a religious community to be granted land on one side of a stream only. To utilise the valley floor for fishponds, therefore, required negotiations with neighbouring landowners. When Byland Abbey decided to construct a new pond in the Wakendale Beck valley in 1234–5, the monks acquired permission to construct a dam and to have a right of way round the western margin of the pond on their neighbour's land for fishing and drawing nets; but if the pond overflowed an adjoining thoroughfare, they would be required to replace the road (McDonnell 1981: 25–26). Construction and management of valley-bottom ponds normally necessitated the diversion of the natural stream into a canalised leat. Such realignments could involve considerable engineering works, as at the Premonstratensian priory of Orford (Everson et al. 1991: 181–183). When the surroundings of the Cistercian abbey of Bordesley were first surveyed in the 1960s it was realised that the entire flow of the River Arrow had been diverted out of its natural bed and canalised along the side of the valley for about 1 km, to create space for two large parallel fishponds and a separate millpond (Aston and Munton 1976); subsequent excavation indicated that the river diversion had taken place before about 1180 (Astill 1993: 104–107).

The creation or enlargement of fishponds usually required the sacrifice of land previously used for other productive purposes, while interference with the natural drainage might cause unintentional flooding nearby. These were frequent causes of conflicts. The Bury St Edmunds chronicler recorded how Abbot Samson (1182–1211)

> Raised the level of the fishpond at Babwell [...] to such an extent that [...] there is no man, rich or poor, having lands by the waterside [...] but has lost his gardens and orchards. The cellarer's pasture on the other side of the bank is destroyed, the arable land of the neighbours is spoiled, the cellarer's meadow is ruined, the infirmarer's orchard is drowned through the overflow of water, and all the neighbours complain of it. Once the cellarer spoke to him in full Chapter concerning the magnitude of the loss, but the abbot at once angrily replied that he was not going to give up his fishpond for the sake of our meadow (Butler 1951: 131).

In the 1270s the Augustinian canons of Cirencester enclosed land at Duntisbourne in order to make a new mill and fishponds, provoking complaints from the Benedictine monks of Gloucester, who claimed that the waters encroached over land where they had rights of common pasture. The Gloucester monks were persuaded to relinquish their claim in 1275 (Devine 1977: no. 367, 641, 653, 655).

Documentation for monastic fishponds declines markedly after the middle of the fourteenth century. Archaeological evidence from Eynsham suggests that pike consumption, reaching a modest peak after construction of the abbey ponds in the early thirteenth century, thereafter declined steadily to the Dissolution (Hardy et al. 2003: 510). It would be easy to dismiss the later middle ages as a time of declining interest in fish production, as monastic landholders, experiencing the same economic difficulties as lay lords, began withdrawing from direct farming, leasing out their ponds with their outlying demesnes. A property leased out by Quarr Abbey in 1430 included a fishpond (Hockey 1970: 177). Winchester Priory's two ponds at Fleet were also leased out by 1491, the prior agreeing to provide the timber necessary for maintenance of the intervening causeway (Currie 1988: 275; 1989: 158). Whereas Ramsey Abbey continued to derive considerable income in the fifteenth century from leasing its fisheries in the rivers and natural meres of the Fenland (Raftis 1957: 300), rents from artificial fishponds were generally low. Winchester Priory asked only 23s 4d per annum for the Fleet ponds, little more than the value of the 100 fish which the lessee was required to send to the priory each year. Low rents probably reflected the expectation that the lessee would provide his own breeding stock, exactly as would be the case for leased pasture land (Currie 1991: 99).

Despite the general move towards leasing, however, fishponds within the precinct and those belonging to manors which had been retained as residences for monastic officials were usually kept in hand. At Titchfield land abandoned after the Black Death had been utilised for a new fishpond before 1393, and both here and at Southwick archaeological evidence confirms that the ponds were not only maintained but also enlarged. New ponds were made at Eynsham before 1360, and ponds at Abingdon and Selby were still producing fish into the fifteenth century (Currie 1989: 154, 158).

Monastic precinct fishponds: layout and form

There is little difference in characteristics of siting, construction, form or management between monastic and secular fishponds in Britain. The layout of ponds within or just outside monastic precincts depended upon a number of factors, of which the most important were the physical characteristics of the site and the amount of space available. In areas of accentuated relief, ponds were confined to chains along narrow, steep-sided valleys, where dams were easily constructed. Where the site was generally level the course of leats bringing in and removing water required more careful surveying, and more earth-moving was needed to excavate beds and provide retaining banks; but there was greater scope to create more elaborate arrangements of ponds for more convenient management. On the whole the reformed monastic orders settling in remote locations had better opportunities to lay out more extensive arrangements of fishponds, simply because they had more space, whereas the older Benedictine abbeys and the later friaries were more frequently in congested urban settings.

Abandoned ponds were easily infilled, and current investigations at Dorchester Abbey demonstrate that present appearance can be a poor reflection of past complexity (R. Weston, *pers. comm.*, 22 October 2012). Nevertheless, it still seems a valid contention that the number and size of ponds within the precinct bears little relation to the wealth or status of the monastery. Some of the grandest Benedictine houses appear meagrely-equipped: Canterbury Cathedral Priory's twelfth-century plan shows only one fishpond within the precinct (Skelton and Harvey 1986: Plate 1B), and only one small medieval pond is evident at Glastonbury Abbey (Burrow 1982). By contrast the small rural Augustinian priory at Maxstoke had eight ponds of various sizes within its precinct, along with a moated enclosure and two more ponds just outside the wall (Holliday 1874). Surveys in Lincolnshire have recorded extensive, complex pond systems belonging to rural religious communities of equally modest status, including the Cistercian nunnery of Heynings, the Gilbertine priory of Sixhills, the Benedictine nunnery of Stainfield and the Premonstratensian priory of Orford (Everson et al. 1991: 112–115, 162–164, 175–176, 181–183). In Hampshire Currie (1988: 270–203) has commented on the sophisticated engineering skills exhibited in the ponds of the Premonstratensian abbey of Titchfield, the Cistercian abbeys of Beaulieu, Quarr and Netley and the Augustinian priory of Southwick, all of which have complex arrangements of leats and sluices permitting independent management of each pond.

Although complex water management schemes have been observed on many monastic sites, their interpretation can be difficult. It has been assumed too readily that any ponds surviving in the immediate vicinity of a monastic house or grange must be fishponds and must be of monastic origin. Ponds also supplied power to mills and had other functions. Around 1300 the Cistercians of Salley Abbey made a drinking-pond for their cattle, 12 m square (McNulty 1934: 44–45, no. 467). Evidence for flax-retting has been noted on several monastic sites in north-western England (Higham 1989: 46–49). At Hulton Abbey a row of four square ponds separated by a long bank from a larger rectangular pond, previously interpreted as fishponds, may be for flax-retting (Klemperer and Boothroyd 2004: 4, 198). Palynological sampling of a presumed fishpond at the Gilbertine priory of Ellerton has demonstrated that its real purpose was hemp-retting (Gearey et al. 2005). Even where sound documentation exists for the presence of monastic fishponds, their identification with surviving water bodies or earthworks usually rests upon circumstantial evidence and presumption rather than scientific proof. Moreover, recent investigations on many sites have shown that we must not underestimate the effect of post-Dissolution conversion of monastic buildings and precincts to domestic use, in particular the extent to which new garden designs involved alterations to monastic pond systems.

Interpretative difficulties presented by field evidence can be illustrated by evolving views on the water system of the Cistercian abbey of Byland. Following pioneering fieldwork in the 1960s, subsequent arguments for increasingly extensive pond systems have since been refined by more rigorous analytical survey.

In 1965 McDonnell and Everest's investigation identified two probable fishponds in the valley immediately north-west of the precinct and two probable millponds west and south of the abbey buildings, all four ponds being fed by a stream flowing down from Cocker Dale to the north-west.

Subsequent reassessments identified six more dams higher up the same valley, though only the uppermost could have retained a pond of any size (McDonnell 1981: 30–31); also two large ponds immediately north of the abbey precinct, later reinterpreted as a single lake (Harrison 1986, 1999); and a further possible fishpond complex west of the uppermost millpond (McDonnell and Harrison 1978); while yet more ponds existed on the granges to the west. If all these identifications were correct, then the Byland monks could have had up to 75 ha. under water.

However, some investigators took a more cautious view, and their doubts have been endorsed by a recent detailed survey by English Heritage. This has suggested, firstly, that neither of the two lowest dams in the valley north-west of the abbey retained permanent ponds. The purpose of the upper dam is now believed to be to divert the flow of water down the valley into the leat which ultimately served the upper mill. The lowest dam, immediately outside the precinct wall, constructed with massive stone blocks on its upstream side, has an asymmetrical profile and lacks any sign of a spillway. No fishbone samples were retrieved from the valley above. This now seems more likely to represent a flood-bank added at a subsequent date to pen back floodwaters on occasions when the dam above was overtopped, in order to prevent excessive flow along the leat damaging the mill below. It may later have served as a causeway, perhaps carrying a realigned precinct boundary (Jecock et al. 2011: 80, 82).

The interpretation of the embankments west and south of the abbey as dams for the fulling-mill and corn-mill has also been called into question. The top of the bank 200 m west of the abbey buildings, the supposed dam serving the upper mill, is not level but slopes down from north to south; it also partly overlies the mill-leat itself. This now seems more likely to represent an incomplete viewing platform, perhaps constructed as late as the eighteenth or early nineteenth century, to provide a picturesque prospect of the abbey ruins. The supposed dam of the lower millpond could only be traced on one side of the stream, and its identity is also now dubious (Pearson et al. 2004: 4–7; Jecock et al. 2011: 66–68, 90).

Finally, although the area of the proposed large pond or ponds north of the precinct is naturally marshy, it lacks any evidence for a dam, and it is difficult to see how this could have been managed as a pond; moreover it contains extensive rectilinear earthworks which suggest that the land had some other use (Jecock et al., 2011: 53, 79–80). The impression that the precinct was almost surrounded by extensive ponds can, therefore, no longer be

upheld, and the closest monastic fishponds now appear to lie in the Wakendale Beck valley two kilometres to the west.

In view of the sheer size and complexity of many monastic pond systems,
it is no surprise that things occasionally went wrong. At Hailes Abbey earthwork survey has identified three pond sites immediately south-east of the
claustral buildings, two of which are shown as retaining water on a 1587 map;
the third is probably the one on which the sluicegate failed during vespers on
the feast of Corpus Christi in 1327 releasing a torrent of muddy water into
the claustral buildings (Coad 1985: 6; Brown 2006: 22–24). At Bordesley
Abbey the monks were unable to sustain their ambitious scheme of water
management in the Arrow valley, which had involved the diversion of the
entire river to accommodate fishponds and a mill: excavation of the mill
showed that it had become choked with flood silt and abandoned well before
the Dissolution (Astill 1993: 104–107).

Small single rectangular ponds, sometimes supplied by leats but often fed
only by local spring seepage, have been noted within numerous monastic
precincts, and can have served only as store-ponds for fish intended for
consumption. Chains of between two and six valley ponds, in which fish
could be bred, are also common, examples including Dunkeswell Abbey,
Eynsham Abbey, Halesowen Abbey, Daventry Priory and Owston Abbey
(Bond 1988: 95–98).

Open level ground permitted more distinctive arrangements of ponds.
Parallel rows of half a dozen or so linear ponds have been noted on several
monastic precincts. This pattern may be a relatively late development, designed
for ease of trawl-netting. On the edge of Kirkstead Abbey's precinct is a group
of ponds of varying length, all linked by a single channel at one end (Bond 2004:
200–201). A similar group in the outer court of Chertsey Abbey (Surrey) lies
within a rectangular area enclosed by further watercourses; here three ponds
are still visible and the remainder, shown on an estate map of 1735, have been
confirmed by geophysical survey; they seem to have been laid out during the
first half of the fourteenth century (Poulton 1988).

More compact groups of small ponds, often enclosed within a ditch or
moat for security, have been recorded within several monastic precincts. The
Benedictine abbey of St Benet's Hulme has a well-preserved group of five
small ponds enclosed within a moat, with three or four parallel ponds
immediately outside it and six or seven more scattered ponds elsewhere
within the precinct. A similar complex of four ponds surrounded by a moat
has been recorded just outside Thornton Abbey's precinct, and there are also
four parallel ponds near Notley Abbey (Knowles and St Joseph 1952: 24–25,

200–201, 222–223). Pond complexes including islands which themselves contain small ponds have been recorded at North Ormesby Priory's grange at North Kelsey and at the Benedictine nunnery of Stainfield (Everson et al. 1991: 139–141, 175–176). Ponds of this type may represent either hatcheries, or sorting tanks, or storage ponds for different types of fish awaiting consumption.

Ponds on some monastic sites were clearly laid out in successive stages. This avoided construction costs peaking at the same time, and made possible adaptation to changing needs. Evesham Abbey's chronicle records that Abbot Randulf (1214–1229) "made a second and third fish pool at Evesham, because there was already one", a statement which accords well with the field evidence and earlier maps (Macray 1863: 261; Bond 1973: 31–32). A survey of Eynsham Abbey made in about 1360 records "fishponds recently-made" in the garden (Salter 1908: 37). It seems unlikely that this refers to the main series of ponds along the Chil Brook valley, which had probably come into existence soon after Abbot Adam's enlargement of the precinct in 1217; a more likely identification is with a group of five or six small ponds, probably store-ponds, slight earthworks of which have been recorded above the valley floor on the southern side of the precinct (Bond 1988: 100–103; Hardy et al. 2003: 509–510).

Fishponds on monastic manors and granges

Records of monastic fishponds outside the precinct are, understandably, most informative for the period of direct demesne management, from about 1200 to 1350. The Evesham Abbey chronicle mentions at least a dozen in nine separate places over this period, and at least five more of the abbey's manors have ponds for which no documentation has been found – some of them possibly made later by lessees (Bond 1973: 29–38). In the late middle ages the accounts of William More, last Prior of Worcester, record his management of ponds attached to his four retained residences at Battenhall, Crowle, Grimley and Hallow. At Battenhall (Figure 1.2) earthworks of four linked ponds surround the moat in the former park, and these can tentatively be matched with the 'Nether Pool', 'Park Pool', Over Park Pool' and 'Dey Pool' mentioned in the accounts (Bond, 2004: 205–206). Aston (1982: 264) has argued that each Battenhall pond could have been regulated at two different levels by use of the sluice gates; but the shallow and deep areas also provided different conditions for feeding and refuge, and helped to concentrate the fish in a more limited space for harvesting when the pond level was lowered. At

Crowle there is no archaeological evidence for a separate fishpond, and the fish were clearly kept in the moat, which survives as an earthwork. Several different ponds are named at both Grimley and Hallow, but cannot be located with certainty (Aston 1982: 258–265).

No artificial ponds came near rivalling the size of certain natural lowland meres exploited as fisheries by monastic communities, such as Whittlesey Mere in the Fenland (Lucas, G., et al. 1998) or Meare Pool in the Somerset moors (Rippon 2004: 119–122). Nevertheless, as Currie (1988: 267–270) has demonstrated in Hampshire, some artificial fishponds on outlying monastic estates themselves achieved considerable size. Boats were needed to fish Winchester Priory's two large ponds at Fleet with nets, and the surviving pond there covers some 53 ha. Beaulieu Abbey's pond at Sowley today occupies about 20 ha., and was originally considerably larger.

Byland Abbey's two fishponds in the Wakendale Beck valley each covered some 18–20 ha. The lower pond, between the granges of Scencliff and Oldstead, was identified by McDonnell and Everest (1965: 37–38), and subsequently another large pond was recognised above it. The upper pond was probably constructed not long after 1147, when the monks were still settled in their earlier location, Stocking. Frictions with a neighbouring landowner forced them to surrender this pond before 1190. The lower pond, constructed to replace it, is dated to the 1230s by consents from more amenable neighbours to the building of the dam and the inundation of part of their land. The new dam was about 400 m long and 9 m high in the centre, utilizing an estimated 45,000 cubic metres of earth, clay and stone. There is some indication that the Byland monks may have regained control of the upper pond by 1315, but no evidence that either pond was maintained after about 1400 (McDonnell 1981: 25–27, 32–33; Burton 2004: 173–176; Jecock et al. 2011: 18–19).

On Cistercian estates elsewhere, the long-running Bordesley Abbey project has included an investigation of the abbey's granges, at least two of which included fishponds. Sheltwood Grange had a chain of three ponds indicated by dams across the valley, and a fishpond is recorded on a lease of 1529. At New Grange a very large millpond occupied the valley bottom, and the tailrace of the mill carried water into an L-shaped arrangement of fishponds on the flank of the valley (Astill et al. 2004: 141–144). Thame Abbey's grange at Shipton Lee in Quainton included a group of ponds immediately west of the farm buildings. Earthwork survey has suggested that there was initially a single pond covering some 1.2 ha. This was subsequently replaced by a more complex arrangement of four ponds covering in total some 0.6 ha., the upper pair simply dug out, the lower pair retained by dams up to 2 m high. The lowest pond may itself have been a later addition. Unfortunately neither the first construction nor the later modifications can be dated firmly to the monastic period, though fishponds are recorded on a manorial survey of 1624. Further west, alongside the River Ray, there are traces of two further ponds covering c.1.8 ha. and 3 ha., the upper pond probably working a mill (Kidd 2006: 150, 153–155).

Stocking and management of ponds

Ponds could be stocked by means of gifts, purchases or transfers of fish within the monastic estate. Henry III allowed a number of abbots grants of live fish from the royal ponds to restock their own fishponds: the Foss Pool at York supplied 60 bream in 1229 to Fountains and 10 prime female bream in 1245 to Byland (Close R. 1227–31: 278; 1242–7: 328), while the royal ponds at Feckenham supplied bream to Evesham in 1240 and to Pershore in 1252 (Close R. 1237–42: 284; 1251–3: 100, 146). The carriage distance from York to Fountains was over 40 kilometres.

Records of purchases for stocking fishponds become more frequent after 1270. In 1291–92 the sacrist of Ely paid 10s for live pike (Chapman 1907: ii, 7). In 1301 the prior of Bicester spent 2s 8d on pike, perch and roach (Blomfield 1884: 142). Abingdon Abbey's pittancer spent 9s 6d on fish to stock the *vivarium* in 1322–23 (Kirk 1892: 3). In 1485 the cellaress of Carrow bought 4,700 roach to stock the ponds within the great garden (Redstone 1946: 69).

Prior More of Worcester's weekly expenditure accounts from 1518 to 1535 provide a particularly valuable record of fishpond management, indicating that some techniques, otherwise undocumented before the seventeenth century, were already routinely practised (Fegan 1914; Hickling 1971; Aston 1982: 258–66). Between 1518 and 1524 Prior More purchased from professional fishermen nearly 6,500 eels for stocking his ponds during the summer: fifty eels could be bought for a penny, and the supplier could provide up to 1,400 eels at a time. Eel purchases ceased in 1524, but some natural restocking clearly continued, since many were found in one of the Hallow ponds in 1534. Restocking with tench, roach, pickerel, bream and perch took place in winter and early spring. This ensured that the fish would be well-grown by the following winter, and could then be fished as needed through summer and autumn. One penny would buy 18 perch, but tench and pickerel could cost up to 3d each. Though contrary to the general medieval practice of autumn stocking, early spring stocking accorded with Taverner's advice that fish intended for store could be caught and handled without damage in cold weather, whereas moving them in warm conditions could be harmful (Taverner 1600: 6). Tench and bream can survive for a considerable time out of water in cold weather, if packed in wet grass or straw. Several orders for store fish were supplied by a fisherman at Ripple, on the River Severn. The overland carriage of live fish from Ripple to even the nearest of More's ponds at Battenhall, 12 km away, would have taken 4–5 hours, while carriage to the more distant ponds could have involved a journey of 8–10 hours; yet such feats were routinely successfully accomplished.

Fairly regular cleansing prevented silting and the build-up of disease. In 1294–95 Westminster Abbey's ponds at Knowle were cleaned out at a cost of £17 4s 11d, an extraordinarily expensive operation (Dyer 1988: 27). When the kitchener of Abingdon Abbey had a pond cleaned out in c.1377 he paid only 8d (Kirk 1892: 40). In 1457, five years after its last cleansing, one 'Geoffrey Dyger and his fellow' were paid 2s 8d for four days' work in cleansing one of Bicester Priory's ponds (Blomfield 1884: 186, 201).

Although Prior More's accounts provide no clear evidence of very regular cleansing, nevertheless each pond was fished out and re-stocked at intervals of four to ten years. This was achieved simply by draining most of the water, bailing out what could not be drained, and netting the fish for stock-taking and redistribution. The larger fish were retained for consumption, the smaller fish sorted for restocking other ponds. The accounts note only 627 large fish being harvested in 19 fishing-out operations taking place over 17 years, an apparently poor return from a recorded stocking of 11,500 fish. However, this figure cannot represent the total yield of all the ponds, as they were clearly fished as required by the prior's own servants, while on two occasions several labourers were hired for this purpose. Every fishing-out also produced large numbers of small fish. One of the Battenhall ponds in 1521 produced 5 cowles of roach and perch (a cowle being a filled container slung on a pole between two men, weighing about 50 kg). Part of this catch was used to stock another of the Battenhall ponds. The moat at Crowle was redug and cleaned out on several occasions through the 1520s, culminating in a major operation in 1533, when sections of the moat were extended to a width of 12 m at the top, 9 m at the bottom; in the following March it was re-stocked with 18 tench and 46 store bream.

Each of Prior More's ponds was kept dry for a period before restocking. This provided an opportunity for other maintenance works. In 1519 a new flood-gate was made for the Nether Pool at Battenhall, costing 3s 3d. In 1528 the dam of one of the Grimley ponds was reconstructed, requiring several loads of clay, thorn scrub and stakes brought in over a period of over three months. Another Grimley pond was equipped with a new floodgate in 1520. Access to the ponds was restricted by hedges and gates. Prior More's journal also records payments for catching predators such as otters and herons.

Fish-houses and associated buildings

Various domestic, processing and storage buildings were associated with medieval fishponds. While these were not restricted to church estates, most of the known references come from monastic records.

The only upstanding medieval 'Fish house' (so-named at least since 1607) is a two-storeyed stone building at Meare, constructed for Glastonbury Abbey in about 1330–1340 (Figure 1.3). This stands alongside the margin of Meare Pool, a natural lake which, before drainage in the early eighteenth century, covered about 250 ha., and was a valuable source of eels, pike, bream and other fish. Several small store ponds were dug close to the fish-house. It had long been assumed that the ground floor of this building was used for the processing and storage of fish and for storing equipment, with accommodation on the upper floor for the abbot's fishermen. However, a recent reassessment by Edward Impey (2009) has questioned both the need here for on-site processing and the suitability of the ground floor for storing nets and equipment. He has argued persuasively that the building was exclusively domestic, the three ground-floor rooms serving as parlour, hall and service rooms, and the two upper rooms as chambers. Although the building is small (6.6 m x 12.36 m externally), its high quality suggests that it served as the occasional residence of a monastic official responsible for managing and protecting the fishery.

Elsewhere equally substantial buildings have not survived. Quarr Abbey's property called 'Fisshehous' on the Isle of Wight was deemed suitable for fortification against invasion from France in 1365 (Hockey 1970: 41, 50, 136). The Durham Priory accounts refer to a glass window being fitted in the fish-house in 1490–1 (Fowler 1898: i, 100). Titchfield Abbey had stone buildings associated with at least two of its fishponds, and a field-name 'Fish House Close' is recorded in the nineteenth century near the dam of Beaulieu Abbey's pond at Sowley (Currie, 1988: 286).

Several humbler functional structures have been identified archaeologically. A tile-roofed timber-framed hut on low sill walls, about 6 m x 4 m externally, overlooked one of Byland Abbey's large ponds near Oldstead

Grange. It was dated by excavation to the later thirteenth and fourteenth centuries, and contained towards one end a large hearth partitioned off from the remainder of the building, suggesting that this was used for fish-smoking rather than domestic warmth. The excavation recovered well over a hundred lead weights, the heavier of which (c. 60–420 g) were thought to belong to seine-nets, the lighter to hand-nets. Quantities of scrap lead suggested that the weights were made on the site. One lead block with remains of an iron ring, carefully chiselled to calibrate it to 7 lb (3.18 kg), may have served to weigh fish for sale (McDonnell 1981: 30; Kemp 1984). A small daub-walled building excavated within a fishpond complex at Washford, believed to have belonged to the Knights Templars, also contained a hearth (Gray 1969). Four separate groups of ponds on Fountains Abbey's Haddockstanes Grange were accompanied by earthworks interpreted as stone-built smokehouses for curing fish (Coppack 1993: 87).

Amenity and symbolism

Monastic ponds were not exclusively utilitarian: those located within gardens often contained fish, but their amenity and symbolic purposes were equally important, if not more so. Prior Wibert's plan of Canterbury Cathedral Priory, drawn up around 1150–60, depicts an oval fishpond (*piscina*) in the south-eastern part of the precinct with twelve semi-circular bays around its margins. Sylvia Landsberg (1995: 37) has suggested that the bays symbolize the twelve apostles, while the central fountain may allude to Christ walking on the waters of Lake Galiliee.

Westminster Abbey's infirmary garden contained one fishpond in 1305–6, and by 1461 there was a second pool. Both were located by augur survey undertaken by the Museum of London in 1988. Payments were made for stocking the first pond with fish and for cleaning out both ponds. However, the infirmarers' accounts contain no record of expenditure on nets or fishing equipment, suggesting that they had limited use as storeponds. They may simply have served as a pleasant garden amenity, where recuperating monks could enjoy recreational angling, while perhaps also providing a habitat for aquatic plants such as water lilies which had medicinal uses (Harvey 1992: 106).

Gardens were often surrounded by ornamental moats, which could also have contained fish. Abbot Godfrey Crowland's new *herbarium* made at Peterborough in 1302 was enclosed within a double moat with four more rectangular fishponds separating it from the river; elements of this complex survived at least to 1721 (Harvey 1981: 12–13, 85, 88–92). Payments for cleaning, repairing and realigning various garden moats are recorded in the gardener's accounts of Norwich Cathedral Priory at intervals from the 1330s through to the 1530s (Noble 1997: 4, 34, 38, 67–8, 73–4). A moated garden may also have existed at the Hospital of St Cross in Winchester (Currie 1992), while the precinct of Waltham Abbey included a moated orchard east of the church and claustral buildings (Huggins 1972: 36).

Excavated monastic fishponds

Relatively few monastic ponds in Britain have been examined by archaeological excavation. Much could still be learned from sections of dams, examination of beds, leats and sluicegate sites, and environmental analysis of waterlogged deposits.

Examination of the dams of two of Titchfield Abbey's ponds showed them to have been constructed with successive layers of rammed material, pure clay in the lowest levels, with more gravel content higher up (Currie 1988: 270). A section at the lower end of Southwick Priory's ponds showed that the original dam, dating from before 1250, had twice been raised between about 1300 and the Dissolution and revetted in timber. A layer of rammed chalk sealed the surface of the earlier dam before the later medieval heightening (Currie 1990: 36–8).

At Taunton Priory a large clay-lined pond edged with wooden stakes and connected with water channels, interpreted as a fish tank, was dated to the first phase of monastic occupation. During a mid-thirteenth century reorganisation of the outer precinct it was infilled with domestic refuse. Fish bones incorporated within the fill were exclusively of marine species (Leach 1984: 111–124, 193–194).

Several ponds at Thelsford Priory were identified by Margaret Gray during emergency excavation in 1972. A small pond which may have served the mill was linked by a sluice channel to a much larger pond to the west, which appears to have been made within an abandoned meander of the nearby brook. In its final form the sluice channel was constructed of dressed stone, but there were traces of an earlier channel bounded by wooden posts. A third small pond lay to the south-east, with a drain connecting it to a small tank; an iron harpoon was recovered from the drain (Chambers and Gray 1988: 130–131). Limited rescue excavation of the Templars' fishponds at Washford recovered a wooden skep with a woven rush base from a possible fish-breeding tank which also contained thirteenth- and fourteenth-century pottery (Gray 1969).

A chain of four fishponds at the small Augustinian priory of Owston survived as a well-preserved series of earthworks into the 1980s. When they were threatened with destruction a 1.4 m by 1.15 m trench was excavated through the deposits in the bed of one of the ponds to sample the environmental evidence. The relative paucity of fish bones and scales may reflect the regular cleansing of the pond and the removal of larger fish for consumption; nevertheless, a residual fish population survived for a while after formal management ceased. Six species were identified, perch, bream, rudd (*Scardinius erythropthalmus* (L.=), roach, pike and chub (*Squalius cephalus* (L.); only chub was regarded as of little food value. The pond was probably abandoned at the Dissolution, and subsequently slowly filled with clay and colluvium over the ensuing four centuries (Shackley et al. 1988).

The monastic contribution to fish farming

Earlier assumptions that monastic houses had played a leading role in the development of fish culture in Britain can no longer be upheld, though they undoubtedly made extensive use of methods previously shown to be 46ffective on lay estates.

Despite the efforts which went into the construction of monastic ponds, reliance upon fish foraging for their own food limited productivity. Although Abingdon Abbey's pittancer spent 2s 2d on feeding the fish in 1322–3, and the abbot's cellarer at Peterborough spent 20d on two sticks of small eels to feed the pike in 1416–17 (Kirk 1892: 3; Greatrex 1984: 24), there is no evidence for supplementary feeding becoming general practice.

It has often been stated that monastic communities led the way in commercial fish farming (Lekai 1977: 217–8; McDonnell 1981: 23). Hockey, believing erroneously that fish were not generally consumed by monks before the late thirteenth century, argued that Quarr Abbey's ponds, if constructed before that period, must have been intended to produce revenue (Hockey 1970: 50). A more recent report from Lincolnshire has suggested that the elaborate ponds on the Templars' preceptory at Willoughton and at the Gilbertine grange of North Kelsey "seem to imply fish-farming on a commercial scale" (Everson et al. 1991: 50–51). The view that monastic fishponds could have a commercial aspect is not entirely without foundation, since surviving gardener's accounts from Abingdon Abbey record receipts from sales of surplus fish on two occasions, 12s 8d in 1388–9, and 21s 4d in 1412–13 (Kirk 1892: 52, 74); but this seems to have been a localised and relatively late development, and there is no evidence that such practices were widespread. Entrepreneurial lessees of monastic manors and granges may have played a more active role in the commercial exploitation of fishponds in the later middle ages, but this requires further investigation.

One important late medieval innovation was the introduction of carp (*Cyprinus carpio* L.). Carp already dominated continental fish-farming by the early fourteenth century, and had been kept in Britain since the fifteenth century; yet no evidence has been found for their presence in monastic ponds before 1530, when the Prior of Llanthony by Gloucester sent a gift of carp to the king (Langston 1942: 136). When Prior More placed "certain carp fisshes" in his Worcestershire ponds in 1531–32, he does not seem to have regarded them as a particular success (Fegan 1914: 345; Hickling 1971: 120–121).

It might have been expected that the centralised organisation of the Cistercian order, with its annual general chapter meeting, would have provided exceptional opportunities for the rapid dissemination of technological advances across Central and Western Europe. However, there is little support for the idea that this had any real effect upon the development and diffusion of fish-farming techniques. Richard C. Hoffmann (1994: 405–406) has reported no evidence from the continent that the Cistercians were involved in fishpond construction or management before the later twelfth century, or that their techniques were any more advanced than those of their lay seigneurial neighbours; and he was sceptical of the Order's effecttiveness in diffusing technical expertise, in view of the delayed adoption of carp by its English and Scandinavian houses (Hoffmann 1994: 405–406, 413 n. 43). David Williams (1998: 365–367) noted that some continental Cistercian houses had over twenty fishponds, but found no reference to their construction before 1146, and no indication even of localised commercial production before the late middle ages. Nevertheless, hints of technical innovation do occasionally appear. In 1346 'pipes and rings' were used in the reconstruction of the fishpond dam at the Cistercian abbey of Croxden (Laurence 1952–53: B.59; Brown and Jones 2009: 43); is this an early reference to a sub-surface conduit removing excess water from the pond, as opposed to the usual above-ground spillway or sluice-gate?

One important advance in pisciculture has been attributed to Dom Pinchon of the French Benedictine abbey of Réome (now Moutiers-St. Jean) in about 1420. Although claims that he bred trout artificially by expressing eggs and milk into a container are not universally accepted, he does seem to have been successful in incubating fertile eggs under sand in wooden hatching boxes (Haime 1854: 1012–1013). However, his discovery remained unpublished for over 400 years, and it was not until the later eighteenth century that renewed experiments with artificial breeding achieved success.

Monastic fishponds after the dissolution

The Dissolution of the Monasteries (1536–40) was accompanied by much plundering of monastic resources. Three hundred carp were removed from the ponds of the suppressed London Charterhouse in 1538, of which 100 went to the king's store (L. & P. Henry VIII 13.ii: no. 903, 375).

Some of the men who rose to positions of wealth and power in the wake of the Dissolution acquired former monastic premises for their private residences and redesigned their gardens and fishpond systems. Thomas Wriothesley, 1st Earl of Southampton, acquired Titchfield Abbey in 1537. Almost immediately he added four new fishponds and acquired 500 carp to stock them, hoping that future sales of carp might yield £20 or £30 a year. The ponds continued to supply carp at least up to 1741 (Currie 1988: 282; 1991: 103). In 1542 Wriothesley further stocked the Titchfield ponds with 1,400 fish, including bream, pike, tench and perch, supplied from Sowley Pond, formerly the property of Beaulieu Abbey (Hockey 1976: 200).

A survey of the site of the Cistercian abbey of Old Warden recorded a substantial dam some 5 m high and over 220 m long, which would have retained a large lake in the valley immediately south-east of the abbey. This was equipped with a diversion channel, and probably is of medieval origin. Ridge and furrow in its bed may indicate a rotation of fish-rearing and agriculture. However, earthworks of a more intricate complex of small rectangular ponds and channels immediately south of the former claustral buildings seem not to be of monastic origin, and were more probably associated with the gardens of the great house built by Sir John Gostwick soon after he acquired the abbey site in 1537 (Baker 1982).

A plan of the site of Leicester Abbey made in 1613 shows four fishponds within the southern extension to the precinct, enclosed around 1500. Geophysical survey has confirmed their position. The gardens were altered by the Hastings family during the later sixteenth century and the ponds may have been inserted or redesigned at this period (Buckley et al., 2006: 14–15, 21, 25). At Bindon Abbey two moated areas with elongated lateral ponds were laid out in the later sixteenth or early seventeenth century when a new manor-house was built within the precinct (Tracy 1987).

Extensive and complex earthworks on the site of the Premonstratensian abbey of Barlings include remains of at least a dozen ponds. Those around the north side of the precinct, now partly destroyed, appear to be of monastic origin; but the long narrow ponds to the south-east adjoin the site of a short-lived seventeenth-century mansion, and probably form part of its garden layout. Similarly, at the Benedictine nunnery of Stainfield, medieval ponds survive with minimal disturbance along the north side of the precinct, but following a reorganisation of the grounds in the early eighteenth century new L-shaped and rectangular ponds were laid out to the east (Everson et al. 1991: 67–9, 175–7).

Excavation at two ponds north of the house occupying the site of the Augustinian priory of Southwick showed that these too had been altered after the Dissolution. The upper pond, a simple rectangular tank equipped with a diversion channel, probably continued in use retaining its monastic form. The lower pond had been enlarged to about 0.4 ha. and compartmented into three, possibly four smaller ponds. This alteration would not have improved the efficiency of the ponds, and evidence of box hedge planting around the lower pond suggests that it was undertaken for ornamental purposes. The lower dam had been raised again between about 1540 and 1750. Documentary records refer to piling and other work on a pond in the late 1690s, though it is uncertain whether this refers to the former monastic ponds or a new pond elsewhere in the grounds (Currie 1990: 36–40).

We can no longer assume all ponds now visible on former monastic sites to be of monastic origin or, even if they are, that they necessarily retain their medieval form. It is becoming increasingly evident that fishponds within monastic precincts were often enlarged and elaborated as an amenity for post-Dissolution residences, some of which themselves had only a relatively brief existence.

References

Aston, M. 1982. Aspects of fishpond construction and maintenance in the 16th and 17th centuries with particular reference to Worcestershire, pp. 257–280 in Slater, T. R. and Jarvis, P. J. (ed.), *Field and forest: an historical geography of Warwickshire and Worcestershire*. Norwich.

Aston, M. and Munton, A.P. 1976. A survey of Bordesley Abbey and its water control system, pp. 24–37 in Rahtz, P. and Hirst, S. (eds.), *Bordesley Abbey*. Oxford. (British Archaeological Reports, British series, 23).

Astill, G. G. 1993. *A medieval industrial complex and its landscape: the metalworking watermills and workshops of Bordesley Abbey*. York (Council for British Archaeology research report 92).

Astill, G., Hirst, S. and Wright, S. M. 2004. The Bordesley Abbey project reviewed. *Archaeological journal* 161: 106–158.

Baker, E. 1982. Warden Abbey. *Archaeological journal* 139: 49–51.

Blomfield, J. C. 1882–94. *History of the Deanery of Bicester* vol. 1–8. Oxford.

Bond, J. 1973. The estates of Evesham Abbey: a preliminary survey of their medieval topography. *Vale of Evesham Historical Society research papers* 4: 1–61.

Bond, J. 1988. Monastic fisheries, pp. 69–112 in Aston, M. (ed.) *Medieval Fish, Fisheries and Fishponds in England*, Oxford (British Archaeological Reports, British series, 182.i).

Bond, J. 2001. Production and consumption of food and drink in the medieval monastery, pp.54–87 in Keevill, G., Aston, M. and Hall, T. (eds.) *Monastic Archaeology*. Oxford.

Bond, J. 2004. *Monastic landscapes*. Stroud (revised edition, 2010).

Brown, G. 2006. *Hailes Abbey and its environs*. London (English Heritage Research Department report series 29/2006).

Brown, G and Jones, B. 2009. *Croxden Abbey and its Environs: an Analytical Earthwork Survey*. London (English Heritage Research Department report series 94/2009).

Buckley, R. et al. 2006. The archaeology of Leicester Abbey, pp. 1–67 in Story, J., Bourne, J. and Buckley, R. (eds.), *Leicester Abbey: Medieval History, Archaeology and Manuscript Studies* (Leicester Archaeological and Historical Society).

Burrow, I. 1982. Earthworks in the south-eastern part of the abbey precinct, Glastonbury. *Proceedings of Somerset Archaeological and Natural History Society* 126: 39–42.

Burton, J. (ed.) 2004. *The Cartulary of Byland Abbey*. (Surtees Society 208).

Butler, H. E. (ed.) 1951. *The Chronicle of Jocelin of Brakelond*. London.

Close R. = *Calendar of Close Rolls*. London.

Chambers, R. A. and Gray, M. 1988. The excavation of fishponds, pp.113–135 in Aston, M. (ed.) *Medieval Fish, Fisheries and Fishponds in England*, Oxford (British Archaeological Reports, British series, 182.i).

Chapman, F. R. (ed.) 1907. *Sacrist Rolls of Ely*, 2. Cambridge.

Coad, J. G. 1985. *Hailes Abbey*. London.

Crawford, O. G. S. 1930. A Saxon fish-pond near Oxford. *Antiquity* 4: 480–483.

Coppack, G. 1993. *Fountains Abbey*. London.

Currie, C. K. 1988. Medieval fishponds in Hampshire, pp. 267–289 in Aston, M. (ed.) *Medieval fish, fisheries and fishponds in England*. Oxford: British Archaeological Reports, British Series, 182.ii.

Currie, C. K. 1989. The role of fishponds in the monastic economy, pp. 147–172 in Gilchrist, R. and Mytum, H. (eds.) *The Archaeology of Rural Monasteries* Oxford (British Archaeological Reports, British series, 203).

Currie, C. K. 1990. Fishponds as garden features, *c.* 1550–1750. *Garden History* 18:i: 22–46.

Currie, C. K. 1991. The early history of the carp and its economic significance in England. *Agricultural history review* 39: 97–107.

Currie, C. K. 1992. St Cross: a medieval moated garden? *Hampshire Gardens Trust journal* 11: 19–22.

DB = *Domesday Book*: among modern English translations, the most convenient is Williams, A. and Martin, G. H. (eds.), 1992. *Domesday Book: a complete translation* London.

Devine, M. (ed.), 1977. *The Cartulary of Cirencester Abbey, Gloucestershire,* vol. 3 Oxford.

Dugdale, W. 1730. *The Antiquities of Warwickshire,* 2nd edition. London.

Dyer, C. C. 1988. The consumption of freshwater fish in England, pp. 27–38 in Aston, M. (ed.) *Medieval Fish, Fisheries and Fishponds in England,* Oxford (British Archaeological Reports, British series, 182.i).

Everson, P. L., Taylor, C. C. and Dunn, C. J. 1991. *Change and Continuity: Rural Settlement in North-West Lincolnshire.* London.

Farrer, W. (ed.) 1914. *Early Yorkshire Charters* vol. 1. Leeds. (Yorkshire Archaeological Society Extra Series.

Fegan, E. S. (ed.) 1914. *Journal of Prior William More.* Worcester: Worcestershire Historical Society.

Fowler, Canon (ed.) 1898. *Extracts from the account rolls of the Abbey of Durham,* vol. 1. Durham (Surtees Society 99).

Gearey, B. R., Hall, A. R., Kenward, H., Bunting, M. J., Lillie, M. C. and Carrott, J. 2005. Recent palaeoenvironmental evidence for the processing of hemp (*Cannabis sativa* L.) in eastern England during the medieval period. *Medieval archaeology* 49: 317–322.

Gray, M. 1969. Worcestershire: Washford. *Medieval archaeology* 13: 283–285.

Greatrex, J. (ed.) 1984. *Account rolls of the obedientiaries of Peterborough.* Northampton (Northamptonshire Record Society 33).

Haime, J. 1854. La pisciculture. *Revue des deux mondes* 2nd series 6: 1012–1013.

Hardy, A., Dodd, A. and Keevill, G. D. 2003. *Ælfric's Abbey: Excavations at Eynsham Abbey, Oxfordshire, 1989–92.* Oxford (Oxford Archaeology: Thames Valley Landscapes, 16).

Harrison, S. 1986. The stonework of Byland Abbey. *Ryedale historian* 13, 26–47.

Harrison, S. 1999. *Byland Abbey.* London.

Harvey, J. H. 1981. *Mediaeval Gardens.* London.

Harvey, J. H. 1992. Westminster Abbey: the infirmarer's garden. *Garden history* 20: 97–115.

Hickling, C. F. 1971. Prior More's fishponds. *Medieval archaeology* 15: 118–123.

Higham, M. C. 1989. Some evidence for 12th- and 13th century linen and woollen textile processing. *Medieval archaeology* 33: 38–52.

Hilton, R. H. (ed.), 1960 *The Stoneleigh Ledger Book.* (Dugdale Society 24).

Hockey, S. F. 1970. *Quarr Abbey and its lands.* Leicester.

Hockey, S. F. 1975. *The Account-Book of Beaulieu Abbey.* Camden (Royal Historical Society, Camden 4th series, 16).

Hockey, S. F. 1976. *Beaulieu: King John's abbey.* London.

Hoffmann, R. 1994. Medieval Cistercian fisheries, natural and artificial, pp. 401–414 in Pressouyre, L. (ed.) *L'espace Cistercien.* Paris.

Holliday, J. R. 1874. Maxstoke Priory, *Transactions of Birmingham and Midland Institute Archaeological Section* 5: 56–105.

Huggins, P. J. 1972. Monastic grange and outer close excavations, 1970–1972. *Transactions of Essex Archaeological Society* 4: 30–127.

Impey, E. 2009. A house for fish or men? The structure, function and significance of the Fish House at Meare, Somerset. *English Heritage historical review* 4: 23–35.

Jecock, M., Burn, A., Brown, G. and Oswald, A. 2011. *Byland Abbey, Ryedale, North Yorkshire: archaeological survey and investigation of part of the precinct and extra-mural area.* (English Heritage Research Department report series 4/2011).

Kemp, R. 1984. A fishkeeper's store at Byland Abbey. *Ryedale historian* 12: 43–51.

Kidd, A. 2006. The Cistercian grange at Grange Farm, Shipton Lee, Quainton. *Records of Buckinghamshire,* 46: 149–156.

Kirk, R. E. G. (ed.) 1892. *Accounts of the Obedientiars of Abingdon Abbey.* (Camden Society new series, 51).

Kitchin, G. W. (ed.) 1892. *Compotus Rolls of the Obedientiaries of St Swithun's Priory, Winchester.* (Hampshire Record Series 7).

Klemperer, W. D. and Boothroyd, N. 2004. *Excavations at Hulton Abbey, Staffordshire, 1987–1994.* (Society for Medieval Archaeology monograph 21).

Knowles, D. 1963. *The Monastic Order in England, 914–1216.* 2nd edition. Cambridge.

Knowles, D. and St Joseph, J. K. S. 1952. *Monastic Sites from the Air.* Cambridge.

L. & P. Henry VIII = *Letters and Papers of Henry VIII.* London.

Landsberg, S. 1995. *The Medieval Garden.* London.

Langston, J. N. 1942. Priors of Llanthony by Gloucester. *Transactions of Bristol and Gloucestershire Archaeological Society* 63: 1–145.

Laurence, M. (ed.) 1952–53. Notes on the chronicle of Croxden, 3. *Transactions of North Staffordshire Field Club* 87.

Leach, P. 1984. *The archaeology of Taunton: excavations and field work to 1980.* Bristol (Western Archaeological Trust excavation monograph 8).

Lekai, L. J. 1977. *The White Monks.* Okauchee, Wisconsin.

Luard, H.R. (ed.) 1865. *Annales Monasterii de Waverleia,* in *Annales Monastici.* London (Rolls Series 36.ii)

Lucas, G., Hall, D., Fryer, V., Irving, B. and French, C. 1998. A medieval fishery on Whittlesea Mere, Cambridgeshire. *Medieval archaeology* 42: 19–44.

McDonnell, J. 1981. *Inland Fisheries in Medieval Yorkshire, 1066–1300.* York (Borthwick Papers 60, Borthwick Institute of Historical Research, University of York).

McDonnell, J. and Everest, M. R. 1965. The "waterworks" of Byland Abbey. *Rydedale historian,* 1: 32–39.

McDonnell, J. and Harrison, S. 1978. Earthworks south of Byland Abbey. *Ryedale historian* 9: 56.

McNulty, J. (ed.) 1934. *The Chartulary of the Cistercian Abbey of St Mary of Salley in Craven* vol. 2. (Yorkshire Archaeological Society record series 90).

Macray, W. D. (ed.) 1863. *Chronicon Abbatiæ de Evesham, ad annum 1418.* London (Rolls Series 29).

Moger, O. M. 1924. Pershore Holy Cross, pp.155–163 in Willis-Bund, J. W. (ed.) *The Victoria History of the county of Worcester* vol. 4. London.

Noble, C. (ed.) 1997. Norwich Cathedral Priory gardeners' accounts, 1329–1530. *Norfolk Record Society* 61: 1–93.

Pearson, T. Ainsworth, S. and Oswald, A. 2004. *An Archaeological Assessment of Earthworks at Byland Abbey, North Yorkshire.* London (English Heritage archaeological investigation reports series AI/33/2004).

Poulton, R. 1988. *Archaeological investigations on the site of Chertsey Abbey.* Guildford. (Surrey Archaeological Society research volume 11).

Raftis, J. A. 1957. *The Estates of Ramsey Abbey: a Study in Economic Growth and Organisation.* Toronto.

Redstone, I. J. 1946. Three Carrow account rolls. *Norfolk archaeology* 29: 43–45.

Rippon, S. 2004. Making the best of a bad situation? Glastonbury Abbey, Meare, and the medieval exploitation of wetland resources in the Somerset Levels. *Medieval archaeology* 48: 91–130.

Salter, H. E. (ed.) 1906–7, 1908. *The Cartulary of the Abbey of Eynsham.* Oxford (Oxford Historical Society, 49, 51).

Saltman, A. (ed.) 1962. *The Cartulary of Tutbury Priory.* London.

Salzman, L. F. (ed.) 1932. *The Cartulary of the Priory of St Pancras of Lewes* vol. 1. (Sussex Record Society 38).

Sawyer, P. H. 1968. *Anglo-Saxon Charters: an Annotated List and Bibliography.* London (Royal Historical Society guides and handbooks 8).

Searle, E. and Ross, B. (eds.) 1967. *The Cellarers Rolls of Battle Abbey, 1275–1513.* Sydney (Sussex Record Society 65).

Shackley, M., Hayne, J. and Wainwright, N. 1988. Environmental analysis of medieval fishpond deposits at Owston Abbey, Leicestershire, pp. 301–308 in Aston, M. (ed.), *Medieval Fish, Fisheries and Fishponds in England.* Oxford (British Archaeological Reports, British series 182.ii).

Skelton, R. A. and Harvey, P. D. A. (eds.) 1986. *Local Maps and Plans from Medieval England*. Oxford.

Taverner, J. 1600. *Certain experiments concerning fish and fruite*. London.

Tracy, C. 1987. Bindon Abbey, pp. 67–68 in Keen, L. and Carreck, A. (eds.), *Historic Landscape of Weld*. Lulworth.

Williams, D. H., 1998. *The Cistercians in the Early Middle Ages*. Leominster.

CJB/CMB 2012

N

Legend:
- ■ Benedictine monks
- ● Benedictine nuns
- ◣ Cluniac monks
- □ Cistercian monks
- ○ Cistercian nuns
- ◇ Carthusian monks
- ▲ Augustinian canons
- △ Premonstratensian canons
- ▽ Premonstratensian sisters
- ◭ Gilbertine canons & sisters
- ◆ Trinitarians
- ★ Military orders
- ⊞ Manors & granges distant from home abbey
- ✛ Other sites

Durham ■

Byland □

Fountains □   York ✛
Salley □   Selby   ▲ Ellerton

Thornton ▲
North Kelsey ⊞   Orford
Willoughton ★   ▽ ▲ North Ormesby
Heynings ○   ▲ Sixhills
   Bullington
   ◭ ● Stainfield
Barlings ◭   Kirkstead □

Hulton □
Croxden □

Tutbury ■   Owston ▲   St Benet Hulme ■
Leicester ▲   Norwich ●
   Carrow ●
   Peterborough ■
Maxstoke ▲   Whittlesey Mere ✛   Ely ■
Halesowen △   Knowle ⊞   Kenilworth   Ramsey ■
Bordesley □   ▲ Stoneleigh
Feckenham ✛ ★ Washford   Daventry   Bury St Edmunds ■
Worcester ■   ◆Thelsford
   ▲ Evesham   ○ Old Warden
Ripple ✛   Pershore
   □ Hailes
Gloucester ■   Bicester ▲
Llanthony ▲   Eynsham ▲   Notley   St Albans ■
   □ Thame   Waltham ▲
Cirencester ▲   Dorchester   London Charterhouse ◇
Abingdon   Westminster ■
   Chertsey ■   Canterbury ■

Meare Pool ✛   □ Waverley
Glastonbury ■   Winchester ■
Taunton ▲   Netley   Battle ■
Dunkeswell □   □ Southwick   Lewes ◣
Beaulieu □ △Titchfield
Bindon □
□ Quarr

km   80
miles   50

Figure 1.1: Monastic fishponds in England. Locations of religious houses and other sites mentioned in the text.

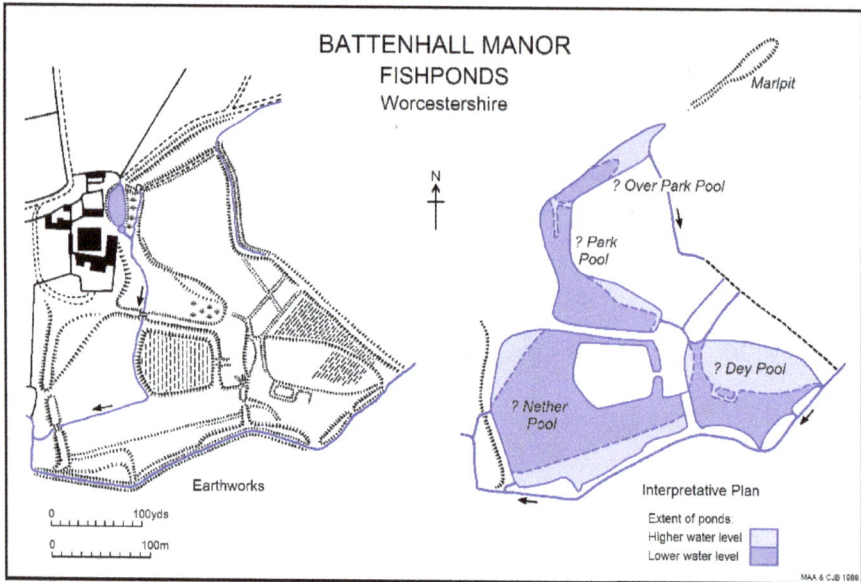

Figure 1.2: Battenhall, Worcestershire. One of several fishpond complexes maintained by William More, last Prior of Worcester, 1518–36.

Figure 1.3: Meare Fish House, Somerset. This is probably the residence of a monastic official supervising Glastonbury Abbey's fishery at Meare Pool.

Figure 1.4: Washford, Warwickshire. This is a fishpond of the Knights Templars. The figure shows the discovery of a fish-skep with a woven rush base during an emergency excavation in 1968.

CHAPTER 2

The History of Aquaculture in Poland

Stanisław Cios

The history of aquaculture with production of carp (*Cyprinus carpio*) in ponds has been well researched in Poland (Brzozowski and Tobiasz 1964; Nyrek 1966, 1970, 1991, 1992; Nyrek and Wiatrowski 1961; Szczygielski 1959ab, 1962, 1967ab; Topolski 1958). The reason for such a strong interest in this topic was the large economic importance of carp culture and abundance of historical sources. The studies on carp culture were published almost solely in Polish. Therefore, in general they remain unknown to international readers. The only publication intended for foreign use was by Chmielewski (1965), who briefly outlined some of the main elements of carp culture.

Very little attention has been given to other fish present in aquaculture. In the mentioned studies, there are just brief passages on the presence of accompanying fish in carp ponds, for example, pike (*Esox lucius*), tench (*Tinca tinca*), Crucian carp (*Carassius carassius*), roach (*Rutilus rutilus*) and perch (*Perca fluviatilis*). However, almost every fish species was present in the ponds because the fish arrived with the inflowing water from the rivers. A study on trout culture from the sixteenth century to 1850 has been published by Cios (2005). This article presents information on three species in aquaculture in Poland: carp, Crucian carp and brown trout.

Carp culture

The origin of carp culture in Poland is still an unclear issue due to lack of sound evidence for the period before 1400. The oldest verbal evidence, concerning Northern Poland, dates back to 1399 (Joachim 1896). From later documents from this region pertaining to the years from 1440 to 1444, it is evident that carp was caught in the Vistula Firth during the whole year, but the number of fish was small (Nowak and Tandecki 1997). The presence of this fish in the firth indicates that it had often escaped from ponds, especially during spates.

As regards the southern part of the country, a family name Carp appears in 1402 in the city of Przemyśl (Smołka and Tymińska 1936). In the years from 1407 to 1414 carp is frequently mentioned in documents from the city of Lwów (Czołowski 1896). In 1424, carp were bought for the king in Bochnia (Krzyżanowski 1909–1913). These sources indicate that carp was well established in this part of Poland at the beginning of the fifteenth century. The striking lack of any reference to carp in the royal accounts from 1388 to 1420 in Kraków and its surroundings, already noted by Hoffmann (1994), may be explained by the fact that at that time carp culture had not yet developed on a mass scale, with a view to supplying Poland's city markets. The fish species (eel, *Anguilla anguilla*, river lamprey, *Lampetra fluviatilis*, catfish, *Silurus glanis*, grayling, *Thymallus thymallus*, salmon *Salmo salar*, roach, *Rutilus rutilus*, bream, *Abramis brama*, sturgeon, *Acipenser oxyrinchus*, burbot *Lota lota*, ide, *Leuciscus idus*, stone loach, *Barbatula barbatula*, and pike, *Esox lucius*) that appear in the accounts were caught in the River Vistula, which flows through Kraków. Apart from carp, there is no reference to any other typical pond species like Crucian carp or tench. The same conclusion may be drawn from the fish species (salmon and pike) often mentioned in other documents from Kraków from the late fourteenth and early fifteenth centuries (Piekosiński and Szujski 1878).

The most plausible hypothesis on the origin of carp in Poland is that it arrived from Germany, possibly also from Moravia. The geographical distribution of the oldest references indicates that the import of carp was probably linked to the German immigration and colonization process in the northern and southern parts of the country, especially in the thirteenth century, when it occurred on a large scale. Such a view would be supported by the findings of remains of carp in Wrocław (Breslau) in layers from the tenth century and twelfth century (Kozikowska 1974; Makowiecki 2003). Carp was either raised there or imported as a foodstuff.

A common term appearing in the Latin location documents from the thirteenth to fourteenth centuries is *piscina*, referring to a fish pond (e.g. *Kodeks...* 1877–1999). The pond was constructed either on the stream (by creating a dam), close to the stream (with a fresh supply of water) or in an open field. It is highly probable that the first carp were raised in Poland in such *piscinae*. *Piscinae* were usually stocked, while *piscaturae* (another old term) were waters with naturally occurring fish.

Carp culture spread quickly in the country in the fifteenth century. The Małopolska region in the southern part of Poland (close to the city of Kraków, the seat of the Polish kings), became the centre of carp production. Large numbers of ponds were constructed in areas near Oświęcim, Brzeszcze, Zator, Pszczyna, Rybnik and Cieszyn. The local topography, hydrological regime and climatic conditions were exceptionally well suited for fish farming. The River Vistula and some of its affluents (especially the Soła) offered excellent opportunities for construction of ponds and a good supply of water. There was also a big market nearby in Kraków, where the fish could be easily transported downstream in large wooden boxes in the water and later sold. Fish farming offered high profits, being one of the main sources of revenue for entrepreneurs. Fish farms were set up by nobility, townsmen, peasants and even the clergy. All of them wanted to participate in this profitable activity.

At that time, the technological breakthrough in carp production consisted of separating spawning ponds from growth ponds. The first ones were small and shallow, allowing the water to warm up quickly and to enable the carp to reproduce. Later on the fry would be transferred to a separate pond, in which they had a considerably higher survival rate than in big ponds. The following spring, they were transferred to big ponds to grow. They would remain there for 3–4 years.

This method of reproducing carp in an artificial environment (in general carp do not reproduce in natural conditions in Poland due to low water temperature) rendered possible obtaining large quantities of fry. They were used for stocking ponds and other water bodies, making this fish accessible even to poorer people.

The system of spawning ponds and successive transfer of fish to different ponds, which permitted more intensive and higher production, was developed in this region. Several other important technical devices were also developed there, like various wooden chutes and water level control systems.

The average production of the ponds was approx. 40–50 kg/ha. In exceptional cases it reached 75 kg/ha. Such a result was achieved without feeding the fish. However, the natural productivity of the ponds was increased through such measures like fertilization and growing plants in dry ponds. Most of the ponds were constructed in areas where the land was not optimal for other types of agriculture, usually in natural depressions, swamps, marshes or stream valleys. Fish ponds therefore increased the total productivity of the land at the disposal of its owner.

The size of ponds was impressive. Many of them were over 100 ha. The largest carp pond still in existence, named after King Sigismundus (this form is used later in the text) Augustus, who ordered its construction in the sixteenth century, covers an area of 460 ha. It is located in north-east Poland.

In response to the large demand for technical knowledge, a manual for carp culture in the Małopolska region was published in 1573 and 1605 by Olbrycht Strumieński (copied by Stroynowski and published in 1609 and 1636). Together with the manual by Johannes Dubravius, originally published in Latin in 1547 (Polish translation in ca. 1660), these two books represented the state of the art in carp culture in central Europe.

Some other regions in Poland, especially in Silesia, also offered good conditions for carp culture. The largest centres were by the rivers Barycz (an affluent of the Oder) and Bzura (Central Poland, an affluent of the Vistula; the latifundia were owned by the Archbishop of Gniezno) and Dniester (Lwów region, nowadays on the territory of Ukraine; formerly famous especially for pike and tench culture). Several ponds were also constructed in the Lublin district. A precondition for development of fish farming in Poland was the presence of a river near the ponds, enabling marketable fish to be transported to the cities.

There were two additional important factors that stimulated the development of carp culture in Poland late in the Middle Ages. The first was the large number of fast days, in some areas even up to 250 days. This required a stable supply of fish produced locally, implying also that fishing in the ponds was conducted on a regular basis (in contrast to modern practice where the crop is harvested at the end of the year, mainly for Christmas, when carp is an obligatory item on the table).

The second factor was the lack of natural lakes in the southern and central parts of the country. Fishing in rivers, though common, could not provide the fish on a regular basis due to natural factors like frequent strong fluctuations in the water level or the presence of ice in winter.

The main fish kept in ponds was carp. However, most other fish species present in Poland could also be found there because they entered the pond with the inflowing water. Among these species, the most highly appreciated and popular ones were pike (according to some data around ten percent of all fish present in ponds; but in 1791, on one estate 17,000 pike were bred, in addition to 18,000 carp, tench and Crucian carp, which were often stocked deliberately. Even some uncommon predators like catfish and pike-perch were stocked there. Smaller fish, especially roach, perch and rudd, were a welcome reward for the peasants who worked for the landlords in the fisheries sector.

The peak in Polish carp culture came in the sixteenth century. The situation changed dramatically with the onset of military activities on Polish territory in the seventeenth century. The frequent passage of armies (especially Polish and Swedish ones) resulted in draining ponds and netting out the fish. Often the fate of the ponds was sealed by the destruction of dikes to speed up the outflow of water. Fish in ponds represented a readily available and abundant food resource for any large unit of ever-hungry soldiers. Though the commanders of the Polish army issued several instructions forbidding looting, including fish ponds, these laws were not obeyed. Hunger was stronger than the law.

There were other social and economic factors that contributed to the decline of carp culture. The first was the relaxation of the strict religious rules concerning fasting. Symptoms of this process were seen in Poland soon after the start of reformation in Western Europe. However, on a larger scale this process was initiated in the second half of the sixteenth century. The first highly symbolic event was a dinner hosted by King Sigismund Augustus in 1548. It was organized during a fast day, but because of the presence of German guests, meat was put on the table. This was an important impulse for considering new ideas on fasting in Poland. The wars conducted in the seventeenth century further contributed to a strong reduction in the number of fast days. The army required strong, healthy soldiers, while fasting was considered to be an obstacle to maintaining their stamina (Cios 2007).

The next factor was the decline in profits in the fisheries sector. Following the economic revolution in Western Europe in the sixteenth century, the international price of wheat soured, quickly changing the economics of agriculture in Poland. It became more profitable to grow and export wheat rather than grow carp, which could not be exported easily. Many ponds were therefore filled in and turned into arable land. The number of professional fishermen quickly dropped (also partly as a result of the ravaging wars). Fewer and fewer people had the knowledge and incentive to farm the fish. Such a situation persisted for 100–200 years, depending on the region.

The reduction in the supply of fish on the market resulted in high prices. Fish, especially the tasty species, later became a luxury that few people could afford. In some areas, money was insufficient to buy fish since even bishops complained about the lack of fish. But the high price no longer acted as a stimulus for economic activity directed towards fish farming. People, especially in the cities, were poorer, so the market was reduced. The inflow of cheap

Russian and German fish after the partition of Poland in 1795 also contributed to the suppression of domestic production. Entrepreneurs' focus was on grain, to which the limited manpower resources were directed.

Changes in pond culture

Until around 1850 there was hardly any technological progress in carp farming. In fact, regression took place in some areas. For example, in some ponds it took six years to grow marketable fish, further decreasing the profitability of fish culture. The successive transfer of fish to different types of ponds was abandoned.

The first symptoms of change appeared in the second half of the eighteenth century. Some enlightened landowners adopted a new attitude towards fish culture. They had seen the potential in raising their incomes through better use of land resources. In response to this growing demand, Kluk (1780) published a book on Polish fisheries with a substantial chapter on pond culture. Other texts on this topic (various instructions and innovations), written under the influence of aristocrats, were also published. The few remaining copies of old books by Strumieński and Stroynowski were frequently copied by hand and manuscripts were circulated among pond managers, who eagerly sought practical instructions.

A major innovation in carp culture was introduced by Thomas Dubisz (1813–1888), a Slovak, who worked in the fish farming sector for many years and in 1869 was charged with the management of carp ponds in the Cieszyn area, owned by Albrecht Habsburg. In a short time, he developed an intensive type of carp culture. Mature carp, in the ratio of two males per each female, were put into shallow and small spawning ponds (100–300 m²). The fry were then transferred to other ponds, where they had better feeding conditions. At the end of the year the fish, which had reached an average weight of 300 g, were transferred to another pond for overwintering. In spring they were transferred to another pond for further growth. By the end of the year, they had attained 800 g and were again transferred to another pond for overwintering. In spring they were released in big ponds, in which they grew to a size of 1500–2000 g by autumn, by which time they were ready for sale.

In a certain respect, this system resembled that of the sixteenth century. The main difference consisted of frequent transfers of fish to different types of ponds suited to the different life stages of carp (in the old days the fish would usually remain in the same pond for a few years). This resulted in a significantly higher yield, around 200 kg/ha., in exceptional cases even up to 360 kg/ha.

Dubisz never published any results of his research. He transmitted his knowledge orally to various students of fish farming. Dubisz made another contribution to carp culture. He carefully selected the fish for breeding and was thus able to create a breed of fast-growing carp that was resistant to diseases. The only important thing he did not know was the composition of the diet of carp. He presumed carp fed only on plants. This was a common opinion until the publication in 1885 of results of research by Jozef Šusta, a Czech, who found that plankton played a crucial role for carp. This was a major discovery and contributed to further progress in carp culture.

Soon afterwards, several other carp producers experimented with carp and achieved significant success. The most well-known of these was Adolf Gasch (1839–1915), who managed several carp ponds in Kaniów, south-west of Kraków. Through careful selection he obtained the famous breed of Galician carp. In 1880 he presented this fish at the exhibition in Berlin, exciting strong interest in his material. Soon he became one of the top experts on carp culture in Europe and a major supplier of carp fry for European markets. After the exhibition in Hamburg in 1883, he also sent fry to Professor Spencer Baird in the United States.

Gasch actively promoted his knowledge in written form. He published several papers promoting the system worked out by Dubisz. Over a short period, his works set the standard for carp culture in Poland.

These developments coincided with another process. In 1879, at the initiative of Maksymilian Nowicki, a professor of zoology at the Jagiellonian University, the National Fisheries Association was created, with its head-quarters in Kraków. The goal of the association was to promote sound management of fisheries according to the latest achievements in science. This initiative was a response to a drastic drop in fish populations in rivers, a result of lack of management, industrial pollution, channelization and destruction of the natural environment. The association promoted scientific research, artificial reproduction of fish, stocking of rivers and lakes, development of carp and trout farming, and adoption of laws governing the fisheries sector. It also strongly promoted angling as a form of recreation in conjunction with proper management of waters. Beginning in 1879, it published a journal (*Okólnik Rybacki*) that was a key means of communication between members of the association, as well as with other people in Poland (and abroad) interested in fisheries.

The association gathered some of the most respected authorities in Poland. Among its members was Ignacy Paderewski, the world famous musician, and prime minister and foreign minister of Poland in 1919. He owned an estate with carp ponds near Kraków. Another member was Julian Nowak, rector of the Jagiellonian University and prime minister of Poland in 1922. He was a well-known fly angler, who contributed several articles on this topic in the association's journal. Other members included an aristocrat, Artur Potocki, and Józef Rozwadowski, the author of Poland's first angling manual, published in 1900. The association thus enjoyed broad political, administrative, financial and intellectual support from the leading personalities in Kraków and the region. It had the capacity to reform fish farming and promote sustainable fisheries in Poland.

The golden period of the association ended in 1918. After WWI, economic and political conditions were different. Formally the association still existed and continued to publish its journal until 1945, but it became just an internal circular with limited impact on fisheries.

The association's activities served as a stimulus for fisheries management in other parts of Poland. Such activities were initiated late in the nineteenth century in Warsaw, Poznań and Vilnius.

Carp was popular because its flesh was highly esteemed in Poland and it was available in most parts of the country. It was consumed, mainly fresh, by both rich and poor. The latter often obtained the fish by poaching in the landlords' ponds, in defiance of the law.

The "tongue" of the carp was considered a delicacy. The oldest references to its high culinary value can be found in seventeenth century poetry (Ochman-Staniszewska 1990, II: 200; Potocki 1696, 710, 1916, II: 486). In the nineteenth century, carp was usually boiled or steamed, with the addition of sweet ingredients (raisins, almonds, honey, and gingerbread). Another old popular way of preparing carp was with a "grey" sauce made from onions (oldest reference in the seventeenth century in Przetocki 1911 no 31) and "black" sauce made from plums (Volckmar 2005, 87; Przetocki 1911 no 30; Potocki 1918, III: 599).

Crucian carp culture

In contemporary Polish literature, this fish is usually portrayed as a less important component of fish stock in carp ponds in the past. According to the data presented by Rybarski (1931: 73), in the ponds of the Duke of

Oświęcim in the first half of the sixteenth century about one per cent of all netted fish, destined for the market, were Crucian carp.

However, from the scanty historical sources available it appears that Crucian carp was the object of deliberate pond culture in Poland. In the sixteenth century, Rej (1914, I: 301) advised stocking some ponds with carp and other ponds with Crucian carp, since this would bring extra profit. Similarly, Zbylitowski (1860: 12) referred in poetry to some ponds having carp and bream, while others had Crucian carp. In one document, it is stated that "when ponds are stocked with Crucian carp, they can be used for the benefit of the castle" (Małecki 1964, II: 26, 43). It was a common practice to catch Crucian carp in some ponds and stock them in others (Tomczak 1963: 47, 113). There were also ponds for "growing Crucian carp" (Ochmański 1963: 59). A pond with Crucian carp was called *karaśnik* (Chłapkowski and Żytkowicz 1992: 131, 277) or *karasiniec* (Tomczak 1963; 47), the name deriving from *karaś* – Crucian carp in Polish. The name Karaśnik today denotes certain small lakes and even a town. It is worth noting that only two fish species have given their name to fish ponds in the past. The other is trout (*pstrąg*), lending its name to *pstrążnik* or *pstrążnica*.

The presence of Crucian carp in ponds was also described by poets in the seventeenth century – by Stanisław Orzelski, who stated that the "pond is full of Crucian carp" (hence probably no other fish species were present, because they would significantly reduce the population of Crucian carp through competition) (Pełczyński 1960: 144) and by Potocki (1918, III: 306), who stated that for a wealthy man there is a big fish from a big pond but for a poorer man a small pond with Crucian carp for a soup.

Of interest is information presented by Fabricius (1802: 66–67), who stated – after Prince Sanguszko – that King Stanisław Leszczyński, introduced the Crucian carp in Lorraine in France in the eighteenth century. Leaving aside the issue of the introduction of this fish in France, which needs to be verified from other sources in France, more important is the concept of stocking Crucian carp. Evidently this source documents the interest of Polish aristocrats in such activities, which must have been fairly common in the country.

Crucian carp was a highly valued fish. Falimirz (1534) preferred it to the carp. In 1471, Crucian carp, together with bream, pike, tench and perch, were bought for the royal couple – "pro prando et cena d. regis et regine" (Grodecki 1951: 380). In another source it is stated that in a pond there are various common fish, "but Crucian carp is the best" (Wyczański 1959: 76). From various other documents from the sixteenth and seventeenth centuries, it appears that Crucian carp was frequently caught for the landlords and the nobility (Oprawko and Schuster 1971, I: 170; Chłapkowski and Żytkowicz 1992, I: 101, 131, 277; Kędzierska 1959: 54). In a letter dated 21 December 1606 written by an aristocrat, Leo Sapieha, to his wife, he states that he had sent her over a dozen Crucian carp, which he had caught personally in Nieśwież (Nâsviž – currently on the territory of Belarus) for her. Nobles catching fish was unusual in Poland since they preferred hunting. In some areas, catching the Crucian carp for the landlords was part of the peasants' serfdom (Ślaski 1916: 137; Sygański 1904: 27, 41).

The two oldest Polish cook books from the second half of the seventeenth century do not shed much light on the role of the Crucian carp for the aristocrats. Czerniecki (2010) mentioned this fish only once, as part of a recipe for several fish species. The cook book of the Radziwiłł family in Nieśwież (Moda... 2011) does not mention the fish. It is known from various other later sources that Crucian carp were usually fried and considered a delicacy.

The main conclusion is that Crucian carp artificial ponds were quite common in Poland. They were small and relatively easy to construct. The reference to small ponds or their construction in location documents from the thirteenth to the fifteenth centuries, mentioned earlier, implies that many ponds were probably intended for Crucian carp. Hence, this was one of the first species in Polish aquaculture.

The scale of such management practices remains unknown. However, it seems that even the aristocracy was interested in possessing such ponds, with a view to a regular supply of fresh fish for fast days. Due to the small size of the ponds, Crucian carp were not destined for the market but for consumption by the owner of the pond.

Most of the information on Crucian carp ponds is from the sixteenth and seventeenth centuries. This seems to indicate that later Crucian carp ponds were less popular. The reason may be a fall in the number of fast days in the eighteenth and nineteenth centuries.

Trout culture

Trout culture was quite common in many areas in Poland from the sixteenth century to the beginning of artificial reproduction in the middle of the nineteenth century. There are over thirty historical sources documenting this. The earliest written account of trout farming by the River Soła dates back to 1532 (Rybarski 1931). However, it seems quite certain that trout culture in Poland was practised much earlier.

The name *Pstrążnica* by the River Sanna already appears in documents from 1257 and 1262. In the seventeenth and eighteenth centuries it was used to denote a trout pond. Thus trout culture appears to be the oldest documented form of aquaculture in Poland.

Trout ponds were mainly in the mountain region in the southern part of the country since the fish were easy to procure there. However, such ponds also existed along lowland and highland streams (e.g. the Wieprz and the Bzura with its affluents) in the central part of the country. In the old days, trout thrived in these waters but later disappeared. This was due to development of agriculture, river channelization and destruction of habitat, especially spawning grounds.

Most of the information indicates that trout farms consisted of single small pond with a small number of fish (a few or a few dozen). In some ponds, there were several hundred fish, since one account states that 175 trout were netted in a single catch (Przyboś 1974, 68).

The ponds were owned by the nobility, who greatly appreciated trout as a delicacy. The fish were therefore usually kept in ponds and caught just prior to consumption, especially before feasts or important occasions. In contrast to carp, pike and tench from big ponds, trout were probably never intended for the market. The demand for them was too high on the part of the landlords.

Most of the ponds were small because it was difficult to provide a large amount of cold and oxygenated water. Trout were usually caught by the peasants, as part of their duties, in streams in the neighbourhood and brought to the pond. Sometimes they were bought, but the price was high. They probably were not kept in the ponds for a long time, at least not for a period of 2–3 years, as in the case of carp. Besides, many of the ponds were constructed close to streams, so during floods the fish would easily escape.

One of the major problems in trout culture was the lack of knowledge of the feeding habits of these fish. In the literature there are instructions to use liver, barley boiled with cow's blood, or even water mixed with wheat flour.

The first information on artificial reproduction of trout in the Polish literature was presented by Waga (1843), following publications in foreign literature. Soon other authors picked up this issue and promoted this new element of fishery management. The first trout hatchery was built in 1850 in Dubie close to Kraków. It was owned by an aristocrat – Adam Potocki. Until the end of the nineteenth century it was the only hatchery in southern Poland conducting artificial reproduction. Eyed eggs were sent to other farms in the region, for further rearing, or small trout were released in streams as part of stocking activities. Late in the nineteenth century, eyed eggs were also imported from Germany and Austria.

Artificial reproduction of trout was also conducted in the Pomerania and Masuria regions (contemporary northern Poland) by the Prussian and Russian authorities at least since the early 1860s. Many of the trout farms produced material for stocking (also salmon, sea trout and whitefish).

At the end of the nineteenth century, rainbow trout was introduced in Poland. The material arrived from Germany. Early in the twentieth century, artificial propagation of this fish was conducted in two hatcheries (Złoty Potok and Dubie) in the southern part of the country. The first import of rainbow trout (50,000 eggs) directly from the United States was in 1924 (Kulmatycki 1924). The next similar consignment was in 1961 (10,000 eggs).

Already in the sixteenth century there were attempts to also keep salmon in ponds. These attempts, however, were unsuccessful; Strumieński had stated that salmon were much more difficult to rear than trout as long ago as 1573.

References

Brzozowski, S. and Tobiasz, M., 1964. Z dziejów rybactwa małopolskiego. *Studia i materiały z dziejów nauki polskiej* 9: 3–102.

Chłapkowski W. and Żytkowicz H. (eds.) 1992. *Lustracja województw ruskiego, podolskiego i bełskiego 1564–1565*. Vol. I. Warszawa-Łódź.

Chmielewski, S. 1965. From the history of freshwater fisheries in Poland. [In:] Freshwater fisheries of Poland. Kraków, pp. 3–13.

Cios, S. 2005. Chów pstrągów w Polsce od XVI w. do połowy XIX w. *Przegląd Rybacki* XXX (4): 47–55.

Cios, S. 2007. *Ryby w życiu Polaków od X do XIX w.* Olsztyn.

Czerniecki, S. 2010. Compendium ferculorum albo zebranie potraw. Edited by J. Dumanowski and M. Spychaj. Warszawa.

Czołowski, A. (ed.) 1896. *Księga przychodów i rozchodów miasta, 1404–1414*. 1896. Lwów.

Dubravius, J. 1547. *De piscinis*. Wrocław.

Dubravius, J. ca 1660. *O rybnikach y rybach, które się w nich chowaią*. Translated by A. Proga. Kraków.

Fabricius, J.-C. 1802. *Voyage en Norwège*. Paris.

Falimirz, S. 1534. *O ziołach y moczy ich*. Kraków.

Grodecki, R. (ed.) 1951. Rachunki wielkorządowe krakowskie z r. 1471. *Archiwum Komisji Historycznej PAU*, 16:365–434.

Hoffmann, R. C. 1994. Remains and verbal evidence of carp (*Cyprinus carpio*) in medieval Europe. pp. 139–158 W. van NEER (ed.) *Fish Exploitation in the Past. Proceedings of the 7th meeting of the ICAZ fish remains working group.* (= Annales du Musee Royal de l'Afrique Centrale, Sciences Zoologiques no. 274). Tervuren.

Hoffmann, R. C. 1995. Environmental change and the culture of Common carp in medieval Europe. *Guelph Ichthyology Reviews* 3: 57–85.

Joachim, E. (ed.) 1896. *Das Marienburger Tresslerbuch der Jahre 1399–1409.* Königsberg.

Kędzierska, Z. (ed.) 1959. *Lustracje województwa rawskiego 1564 i 1570.* Warszawa.

Kluk, K. 1780. *Zwierząt domowych i dzikich, osobliwie kraiowych, historyi naturalney początki i gospodarstwo.* Vol. III. O gadzie i rybach. Warszawa.

Kodeks dyplomatyczny Wielkopolski. 1877–1999. Vol. 1–11. Warszawa-Poznań.

Kozikowska, Z. 1974. Ryby w pokarmie średniowiecznych (X-XIV w.) mieszkańców Wrocławia na Ostrowie Tumskim jako wskaźnik gatunków łowionych w wodach danych okolic lub docierających tam drogą handlu. Acta Universitatis Wratislaviensis, 223, *Prace Zoologiczne* 6: 3–14.

Krzyżanowski, S. (ed.) 1909–1913. Podwody kazimierzowskie 1407–1432. *Archiwum Komisji Historycznej PAU* 11: 392–465.

Kulmatycki, W. 1924. Pierwszy transport ikry pstrąga tęczowego z stanów Zjednoczonych Ameryki Północnej do Polski. *Przegląd Rybacki* 5: 253–258.

Makowiecki, D. 2003. *Historia ryb i rybołówstwa w holocenie na Niżu Polskim w świetle badań archeoichtiologicznych.* Poznań.

Małecki, J. (ed.) 1962–1964. *Lustracja województwa krakowskiego 1564.* Vol. I–II. Warszawa.

Moda bardzo dobra smażenia różnych konfektów i innych słodkości, a także przyrządzania wszelakich potraw, pieczenia chleba i inne sekreta gospodarskie i kuchenne. Dumanowski J., and Jankowski R. (eds.), Monumenta Poloniae Culinaria II. Wilanów.

Nowak, Z. H. And J. TANDECKI (eds.). 1997. *Księga rachunków urzędów rybickich komturstw malborskiego i dzierzgońskiego 1440–1445.* Toruń.

Nyrek, A. 1966. Gospodarka rybna na górnym Śląsku od połowy XVI do połowy XIX wieku. *Prace Wrocławskiego Towarzystwa Naukowego,* ser. A, 111. Nyrek, A. 1970. Problem upadku i odrodzenia gospodarki stawowej w Polsce i Czechach w XIX wieku. Acta Universitatis Wratislaviensis, 109, *Historia* 17: 131–142.

Nyrek, A. 1991. Przejście od chowu do hodowli ryb na Śląsku w XIX w. [In:] *Studia nad rybołówstwem w Polsce.* Toruń, pp. 85–99.

Nyrek, A. 1992. *Kultura użytkowania gruntów uprawnych, lasów i wód na Śląsku od XV do XX wieku.* Acta Universitatis Wratislaviensis, 1361, Historia, 97.

Nyrek, A. & Wiatrowski, L., 1961. Gospodarka rybna w księstwie pszczyńskim od końca XVII do początku XIX w. Zeszyty Naukowe Uniwersytetu Wrocławskiego, ser. A, 30, *Historia* 5: 81–127.

Ochman-staniszewska, S. (ed.) 1989–1991. *Pisma polityczne z czasów panowania Jana Staniszewska Wazy 1648–1668.* Vol. 1–3. Wrocław.

Ochmański, W. (ed.) 1963. *Lustracja województwa sandomierskiego 1564–1565.* Wrocław-Warszawa-Kraków.

Oprawko, H. and K. SCHUSTER (eds.) 1971. *Lustracja województwa sandomierskiego 1660–1664.* Vol. I. Kraków.

Pelczyński, M. 1960. *Studia macaronica. Stanisław Orzelski na tle poezji makaronicznej w Polsce.* Prace Komisji Filologicznej, Poznańskie Towarzystwo Przyjaciół Nauk, Wydział Filologiczno-Filozoficzny, 20(1).

Piekosiński, T. (ed.) 1896. *Rachunki dworu króla Władysława Jagiełły i królowej Jadwigi z lat 1388 do 1420.* Kraków.

Piekosiński, T. & Szujski, J. (eds.) 1878. *Najstarsze księgi i rachunki miasta Krakowa od r. 1300 do 1400.* Kraków.

Potocki, W. 1696. *Poczet herbów szlachty korony polskiey y Wielkiego Xięstwa Litewskiego.* Kraków.

Prochaska, A. (ed.) 1892. *Archiwum domu Sapiehów.* Vol. 1. Lwów.

[Przetocki, H.] 1911. *Postny obiad abo zabaweczka wymyślona przez P.H.P.W.* Kraków.

Przyboś A. 1974. *Podróż królewicza Władysława Wazy do krajów Europy Zachodniej w latach 1624–1625 w świetle ówczesnych relacji.* Kraków

Rej, M. 1914. *Zwierciadło.* T. Vol: 1–2. Kraków.

Rybarski, R. 1931. *Gospodarstwo Księstwa Oświęcimskiego w XVI wieku.* Rozprawy Wydziału Historyczno-filozoficznego PAU, ser. II, 43(2).

Smołka, J. & Tymińska, Z. (eds.). 1936. Księga ławnicza 1402–1445. Przemyśl.

Strumieński, O. 1987. *O sprawie, sypaniu, wymierzaniu i rybieniu stawów, także o przekopach, o ważeniu i prowadzeniu wody.* Opole.

Sygański, J. 1904. *Arendy klasztoru starosandeckiego w XVI. i XVII. wieku.* Lwów.

Szczygielski, W. 1959a. Technika gospodarki stawowej w Wieluńskiem w XVI–XVIII w. *Łódzkie Studia Etnograficzne* 1: 21–39.

Szczygielski, W. 1959b. Dzieje gospodarki stawowo-rybnej w wieluńskiem od XVI do końca XVIII wieku. *Rocznik Łódzki* 2: 227–247.

Szczygielski, W. 1962. Rola gospodarki stawowej w życiu ekonomicznym Polski XVI wieku. *Zeszyty Naukowe Uniwersytetu Łódzkiego, Nauki Humanistyczno-społeczne, ser.* I, 27: 47–68.

Szczygielski, W. 1967a. Z dziejów gospodarki rybnej w Polsce w XVI–XVIII wieku. Studia i materiały z historii kultury materialnej, XXXI, *Studia z dziejów gospodarstwa wiejskiego* 9(2): 1–247.

Szczygielski, W. 1967b. *Zarys dziejów rybactwa śródlądowego w Polsce.* Warszawa.

Ślaski, B. 1916. *Materjały i przyczynki do dziejów nadmorskiego miasta Pucka oraz dawnej ziemi Puckiej.* Warszawa.

Tomaczak, A. (ed.) 1963. *Lustracja województw wielkopolskich i kujawskich 1564– 1565.* Vol. II. Bydgoszcz.

Topolski, J. 1958. *Gospodarstwo wiejskie w dobrach arcybiskupstwa gnieźnieńskiego od XVI do XVIII wieku.* Poznań.

Volckmar, M. 2005. *Nicolausa Volckmara Viertzig Dialogi 1612, źródło do badań nad życiem codziennym w dawnym Gdańsku.* Gdańsk.

Waga, A. 1843. O praktycznym sposobie wyhodowania łososia i innych ryb na gospodarski użytek. *Biblioteka Warszawska,* 4: 493–500.

Wyczański, A. (ed.) 1959. *Lustracje województwa lubelskiego 1565.* Wrocław-Warszawa.

[Zbylitowski, A.] 1860. *Niektóre poezye Andrzeja i Piotra Zbylitowskich.* Kraków.

Figure 2.1: The old emblem of the Korzbog noble family with three carp.

Figure 2.2: Pond in Brzezany (modern day eastern Ukraine).
Postcard dated ca. 1930.

Figure 2.3: The Bagiennik pond. (Photo: Andrzej Cios, September 2012)

Figure 2.4: Empty pond for freshly hatched fish. Such a pond is filled with water in May. Carp fry are put into it in late May or early June, when the water temperature is highest. They remain in the pond for a few weeks. Afterwards they are transferred to a larger pond. The pond is then drained and plants are grown. The following year the production cycle of the pond will begin again. Plants also serve as a refuge for carp fry, protecting them from bird predation. (Photo: Andrzej Cios, September 2012)

CHAPTER 3

From Carp to Rainbow Trout
Freshwater Fish Production in Denmark
(written 2012)

Erik Hofmeister

Fresh water fish have been produced in Denmark for many centuries. Fish farming is employed to overcome the seasonal fluctuations and uncertainty inherent in traditional fishing for native species. Aquaculture has transformed the fundamental nature of Danish fishing by introducing new fish species and forms of production, although there are significant gaps in the historical knowledge of Danish fish farming. The older fish farms, which mainly produced Common carp (*Cyprinus carpius* L.) and Crucian carp (*Carassius carassius* (L.)), were a nationwide phenomenon. Modern trout farming – which began at the end of the 1800s – is however primarily concentrated to Jutland, which has the most suitable streams for trout production. Although fish farming was traditionally targeted for domestic consumption, trout farming quickly became a strong export industry.

We do not know exactly when and how aquaculture production originated in Denmark. This topic has not been thoroughly researched and due to the scarcity of written sources and archaeological material, it is unlikely we will ever know with certainty the origins of Danish aquaculture production. Having said this, there is no doubt that the first steps in fish production were taken in the Middle Ages, during which period fish played a major role.

Fishponds are first mentioned in the Danish Landscape Laws. The Law of Jutland (Jyske Lov) mentions 'fishpond' in 1241 (First Book, Chapter 57), the Law of Scandia (Skånske Lov) refers to water which is "guaranteed by the dam" and which can be fished (Chapter 211), and Erik's Law of Zealand (Eriks Sjællandske Lov) mentions 'dug ponds' and 'fishponds' and – in the third book

(Chapter 7) – fish water in which a man has "set fish out". The latter citation shows that the release of fish in artificial ponds – whether dredged or dammed – took place in the High Middle Ages (Rasmussen 1959).

Medieval title deeds and letters often list 'fish water' or 'fish farms' and 'eel farms' among the goods of the estate (Møller 1953). These were dedicated facilities where wild fish could be caught in the watercourse. There are fewer references to "fishponds" where fish could be farmed but some examples do exist. For instance, Absalon Andersen bequeathed land "with woods and fields, meadows [...] fishponds and mill and the other adjoining land of eternal charity" to the Sorø Monastery in 1284. An exchange in 1314 between Antvorskov Abbey on Zealand and Roskilde Church resulted in the abbey receiving all church property in Ramsømagle with "meadows, pastures, fishponds". An exchange in 1465 between two priests in Roskilde included a 'fish park'.

The sources all mention 'fishponds' but we know nothing about the extent, nature or organization of these fishponds. Nor do we know whether the ponds were used for actual fish production or to capture or keep wild fish for consumption.

There is little information about which type of fish may have been farmed. According to the Aarhus Canon Book of Records (Århus Kannikebords Jordebog) from 1315, which itemises the church estates, the church on Helgenaes had two ponds in Borup, "in which there are some good fish". But the book contains no information about which species of fish they were. They were most probably naturally occurring fish which were easy to breed, such as Crucian carp or tench. A 1498 financial report to Duke Frederik in Southern Jutland – later King Frederik I – refers to a "Crucian carp pond" (*karussedam*), which indicates that Crucian carp was employed as a pond fish in the Late Middle Ages (Hofmeister 2004).

The Danish monasteries and carp

Some literature assumes that Danish monasteries introduced the Common carp as an aquaculture fish and subsequently the captive breeding of carp spread nationwide. This assumption is justified by the fact that the strictest monastic orders prohibited the consumption of mammalian meat and fish thus became a crucial food source for the internationally-oriented monasteries, which are assumed to have imported the practice of Common carp farming from abroad. Monasteries in both southern Europe and England had carp ponds.

There is no doubt that Danish monasteries had fishponds in the Middle Ages. In 1486, Pope Innocent VIII confirmed the Voer Monastery's real-estate holdings, which included the deserted Vissing Monastery's possessions with "streams, fishponds and rights" and financial statements from Skandeborg from 1649 mention the Ring Monastery's "three ponds", which probably date back to medieval times. Skov Monastery (present day Herlufsholm School) had several fishponds, one of which, the "Great Park", can still be seen today. Monasteries in Borglum and Mariager and the Marie Monastery in Knardrup are also believed to have had fishponds (Kristensen 2004, 2013).

However, there is a lack of solid evidence that the monasteries introduced common carp farming to Denmark. No common carp bones dating back to medieval times have been found in the country. Archaeological excavations at Øm Monastery in Jutland have uncovered numerous fish bones but no common carp bones. On the other hand, bones of lake trout, perch, pike, roach, bream, crucian carp, eel, garfish, cod, haddock and flounder have been discovered. (Garner 2002)

National and Royal Court financial statements from the mid-fifteenth into the sixteenth century, for example, The Royal Holdings Accounts of Queen Christine (1426–1521) list payment for fish on several occasions but do not mention common carp. Payments are however listed for eel, perch, eelpout, bream, flounder, pike, crucian carp, ling, salmon, mackerel, ray, lamprey, herring and cod. Around 1500, Duke Frederik had a large fishery at Gottorp Castle in southern Jutland, which included fishponds at the castle. The freshwater fish included eel, perch, bream, pike, crucian carp and river lamprey – but no common carp.

As it stands now, there is no reliable evidence of the existence of common carp in medieval Denmark. However, it is certain that common carp farming occurred on Danish soil in the late sixteenth century.

Fish farming in the early modern period (1500–1800s)

While there are many gaps in our knowledge of the medieval fish farms, we are on safer ground when it comes to fish farming operations from the mid-sixteenth and the following centuries. Financial statements, leases and correspondence shed light on the fish farming operations of royalty and the nobility, who were the major backers of fish farming.

Aquaculture entered a new phase during this period with the introduction of the common carp as a breeding fish. The common carp, a native Asian fish, was brought to Europe by the Romans. It was introduced to central Europe in the early Middle Ages and subsequently spread further north. The common carp is a very meaty, tasty fish with a high fat content. It requires meticulous care and favourable water and bottom conditions. But other fish, including the substantially smaller crucian carp (*Carassius carassius*), were also farmed. Unlike the common carp, the crucian carp is extremely hardy and thrives in large quantities even in less favourable environments. To a lesser extent, tench (*Tinca tinca*) was also farmed, as was the small stone loach (*Barbatula barbatula*), which we know was farmed abroad (cf. Lundberg and Svanberg 2010). Financial statements from Skanderborg from 1649/50 include a 'stone loach pond' (*Smerlingdam*) and there is a later reference to a 'stone loach pond' on royal property in Hillerød. When a new royal fish master took office in southern Jutland in 1655, he was assigned 100 stone loach in addition to common and crucian carp (Kristensen 1973). But the common and crucian carp were of primary importance in aquaculture at the time.

Common carp were farmed on the Faurholm Estate in Frederiksborg from 1560, but the best known propagator of common carp farming was the landowner, Peter Oxe, who during his many trips aboard became acquainted with the well-organised common carp farming in Central Europe. In 1566, Peter Oxe brought some brood stock from France to his estate in Gisselfeld, where he established several common carp ponds. Peter Oxe's initiative was quickly rewarded with success and by the year 1600, common carp farming was in full bloom.

It should be noted that when common carp farming flourished in the late 1500s, salt water fishing in Denmark began a decline that intensified in the 1600s. Freshwater fishing, both traditional and farmed, became increasingly important to provide a relatively steady supply of fish. Another contributing factor was that many landowners in the late sixteenth century sought to optimise the operation of their estates (Bager 2000a, 2000b).

Fish farming took place in ponds – artificial ponds – which, singly or in groups of 2–3, were constructed around the castles and outbuildings. The ponds – which were surrounded by dams – were reinforced by palisades and could be drained of water.

Some lakes were also used to raise carp, for example Ejlemade Lake on Bregentved Estate and Skjoldenaesholm Estate Lake, which was used as nursery waters for two-year-old carp. In order to reap the full benefit of fish production, the lakes could be drained of water.

The royal fish farms

Farmed fish operations, mainly common and crucian carp, were widespread among the nobility in the seventeenth century, but the royal house also paid close attention to the new aquaculture developments. Christian IV issued a regulation in 1615 for the royal fish masters responsible for the fisheries in Frederiksborg and Copenhagen. Therein he stated, among other things, that fish masters exercise "diligent supervision" of the ponds so they did not deteriorate and that carp should be treated "without negligence".

Lake fishing, and especially fish farming operations, required technical knowledge and experience. From the outset, the King summoned experienced fish masters from Germany, where aquaculture had been widespread for centuries.

During the seventeenth century, the royal fish farm in northern Zealand developed to become the most extensive in the country. This is not least due to the fact that from the time of Frederik II's reign, northern Zealand became the royal family's preferred residential area and the crown had very large estates in the region. In addition, the area contained numerous large and small lakes which could satisfy the royal court's very strong demand for freshwater fish.

Prior to 1653, freshwater fisheries – including fish farms – in the three northern counties on Zealand (Kronborg, Frederiksborg and Copenhagen) were managed by several fishery managers, who might have other orders on the side. But in 1653 Frederik III decided to lease fishing operations to one man rather than divide the responsibility among several (Rockstroh 1913; Bager 2002).

Under the new scheme, it appears that the first fishery manager was contractually obliged to supply 60 kilogrammes of freshwater fish to the royal court daily throughout the year regardless of whether it was currently situated in Copenhagen or elsewhere on Zealand. The fishery manager was to follow the royal court throughout Zealand with his gear in order to constantly provide fresh fish, travelling from his headquarters at Frederiksborg Castle. In total, he had to deliver 22 tonnes, a considerable quantity, of fresh fish annually to the royal court.

In addition to the royal fish farming operation on Zealand, fish were farmed on other royal properties. For example, according to Skanderborg's financial statements there were fishponds at Skanderborg Castle in Jutland in 1609–10 and 23 fishponds were constructed or restored at the castle in 1618.

There were also royal aquaculture operations in southern Jutland. When a new royal fishery manager took office in 1655, he assumed his predecessor's fish stock, but the king commanded him to purchase more fish so that he had a total inventory of 12,620 common carp, 18,880 crucian carp and 100 stone loach. The fish were placed in a large number of ponds in Haderslev.

Destruction and prosperity

Although aquaculture was still progressing in the 1600s, there were setbacks. Pestilence, hard frosts, and war could be devastating to fish farming operations. The winters of 1654–5 and 1657–8 resulted in major losses of fish due to severe frosts. The winter of 1708–9 was also very harsh and it is believed that 60,000 common carp died in three Zealand counties as a result of frost.

The Danish-Swedish wars of the seventeenth century were also a major setback to fish farming operations throughout the country. Ponds were emptied of fish and often destroyed. The historian K.C. Rockstroh (1913) has vividly described the situation: "But worse when the Swedes came back in the summer of the same year (1658) and remained on Zealand for nearly two years, for the Swedes were very eager for fresh fish. They emptied one fishpond after another and ate the fish, but forgot to pay for it – yes even destroyed dams and line works, partly for an easy way to get hold of the fish and, on the other hand, a desire for destruction." Southern Jutland was also angry. All ponds in Haderslev County were destroyed in 1657 and there were no fish left in the ponds and lakes after the war (Kristensen 1973).

However, the ponds were restored and during the eighteenth century the royal and aristocratic carp farming operations reached their peak domestically. There were sizable pond systems in some places. In 1682, the royal fish farming operation in Frederiksborg County consisted of sixteen ponds and Kronborg County had fourteen ponds. Nearly a hundred years later, in 1769, the number of common and crucian carp ponds in Frederiksborg County had increased to 76 and Kronborg County had 25 ponds of common and/or crucian carp. There were 84 lakes, streams and ponds on the Hørsholm estate in 1752, including 24 dedicated ponds with an inventory of 35,200 common carp and 38,400 crucian carp in addition to the many thousands that were in extended ponds and breeding ponds (Rockstroh 1913; Bager 2002).

The nobility could also take part. On nobleman Peder Oxe's estate, Gisselfeld, his successors also farmed carp, an operation that peaked around 1700, at which time the fishery included three natural lakes and 26 artificial ponds with carp and tench.

Decline of common and crucian carp farming

The number of common carp and crucian carp fish farms decreased at the end of the eighteenth century and continued to fall throughout the nineteenth century due to declining demand. We can see the fall in numbers in 1783 when the royal fishery manager in northern Zealand reported to the royal fishery administration, which was established in 1781, when there were only 55 working ponds in Frederiksborg County compared to 76 in 1752. The number of fish farms in Kronborg County was reduced accordingly. The following decades experienced a continued gradual decline in the number of royal fish farms, due partly to dehydration and lack of maintenance, and partly because some were sold to private individuals. The noble estates also declined. The Gisselfeld Estate had only 16 small ponds remaining.

There was an attempt to compensate for the declining demand by exporting live carp, especially to Germany. In 1914, there were only about 10 common carp farms remaining, mostly on Zealand. Approximately 25 tonnes of live carp were exported to Germany at this time. Seven tonnes of common carp were exported to Germany in 1926. High tariffs in the 1930s resulted in an attempt to export to other countries, particularly England.

An increasing number of carp farms were closed. Only a few estates, including Gisselfeld and the neighbouring Bregentved Estate, sought to maintain carp production through improved efficiency. But in 1962, even Gisselfeld, after having been in continuous operation for nearly 400 years, ceased carp farming entirely.

A similar picture emerges on another estate, Gråsten Castle, which produced 2.5 tonnes of carp annually during World War II. But stagnating sales of carp, which began in the 1960s, caused the entire operation to grind to a halt in 1980. One can still buy carp today at Gråsten Castle, which adjoins and is part of the Gråsten State Forest, where carp farming resumed in 1990 in order to maintain a special piece of cultural history.

Modern fish farming

At the end of the 1800s, a completely new type of fish and fish farm – the trout farm – was experimented with and a new phase of Denmark's aquacultural history began. The first trout farm was established in 1894 and we believe there were already 50–60 trout farms in Denmark in 1907. Some trout farms were established by farmers as a side business, while other larger farms were founded by limited companies or partnerships. To begin with, they experimented with Danish trout and American brook trout, but these were soon supplanted by the rainbow trout, which was imported to Europe from North America. Rainbow trout were robust and grew rapidly.

Trout farms consisted of several earthen ponds that were excavated by streams where the water could be dammed and diverted to the ponds and then drained when desired. It was relatively cheap to build these ponds and in addition there was plenty of fodder in the form of industrially fished saltwater fish (herring, whiting, dab, etc.). The trout farms were never far from a fishing port, so the feed fish could be chopped into small pieces and fed to the trout. The country thus had ideal locations to establish and operate trout farms and Denmark quickly became the leading producer of trout in Europe.

In the first phase of trout aquaculture history, Germany was the largest purchaser of Danish farmed trout and fish export companies were quickly established. World War I hampered trout sales and after the war sales rebounded slowly. Trout sales did not truly recover until the end of the 1930s. In 1934, 662 tonnes of trout were exported and in 1939 trout exports had increased to 1,148 tonnes. New countries began importing Danish trout, including England (which was the largest importer in the late 1930s), Belgium, Italy and Sweden. In order to meet the increasing demand, Danish trout farmers established many new trout farms in the early 1930s (Hesel 1993; Hofmeister 2004).

The Danes' taste for freshwater fish

The Danes taste for freshwater fish has changed markedly over the years. Today, freshwater fish plays virtually no role in the daily diet. A 2002 survey found that the best-selling fish in Danish fish shops was salmon (although a freshwater fish, the salmon sold in Denmark is harvested from the sea), plaice, cod, herring and so on. The first true freshwater fish on the list is farmed rainbow trout, which comes in tenth place (Dahl 1990; Fritzbøger 2004).

The situation was different in the past when Danes ate a larger quantity, as well as a larger variety, of freshwater fish. There are stories of employees on farms near large rivers who were ensured that they could only be served salmon once a week. In the late Middle Ages, freshwater fish was among the most expensive foods. From 1500 to 1800, the wealthy ate many freshwater fish and major feasts featured freshwater fish as a central part of the menu. There was a distinction between 'gentleman's fish' and 'table fish'. The term 'gentleman's fish' was used as a comprehensive designation for common carp, crucian carp, pike and perch which, according to the taste of the times, were considered better and finer than any other fish, which were called 'mediocre fish' (Bager 2002).

In different parts of the country, shellfish played an important role as food for the poor. It was said in the town of Randers in the 1800s that "no shellfish were on the deserted streets of Randers". Shellfish meant the difference between life and death for many poor people in Randers.

The Danes' evolving taste for fish is reflected in various cookbooks. In 1649, Jørgen Holst published the first fish cookbook in Danish, *Oekonoma Nova paa Danske*, with serving instructions for a variety of Danish freshwater fish. In the first edition of *Frøken Jensens Kogebog* 'Miss Jensen's Cookbook' published in 1901 – a classic cookbook for ordinary Danish cuisine – there are several recipes for freshwater fish, including trout, perch, bream, tench, eel and carp. In recent editions, the number of these recipes has been reduced and recipes for tench and bream are now completely omitted. Of the 22 recipes for fish and shellfish in "A Heartfelt Good Life" (1977) published by the Heart Association, there is not a single recipe for freshwater fish. In an attempt to generate interest in freshwater fish, "A Gastronomic Pleasure Cruise along the Gudenå River" was published in 1996, which featured a variety of recipes for fish that could be caught in the river, including walleye, salmon, pike, tench, perch, eel, whitefish and brook trout (Balle and Mikkelsen 1996).

References

Bager, M. 2000a. Ferskvandsfiskeriet i Søhøjlandet 1570–1640. *Nyt fra Ferskvands-museet* 11: 2–9.

Bager, M. 2000b. Dansk fiskeri i 1500- og 1600-tallet. *Sjæklen. Årbog for Fiskeri- og Søfartsmuseet*. 2000: 27–33.

Bager, M. 2002. Herrefisk og spisefisk fra de nordsjællandske søer, pp. 49–62 in H. Goldsmidt and S. Braagaard (eds.) *At leve med de ferske vande – dengang, nu og i fremtiden*. Silkeborg.

Balle, L and Mikkelsen, M. 1996. Fisk fra Flod til Fad: en gastronomisk naturvandring langs Gudenåen. Silkeborg.

Dahl, J. 1990. Ferskvandsfiskeri gennem de sidste 200 år, pp. 49–62 in *Vandløb og kulturhistorie: rapport fra et seminar afholdt på Odense Universitet den 16.-17.*

januar 1990. Odense: Odense Universitet (= Skrifter fra Historisk institut, Odense Universitet 39).

Fritzbøger, B. 2004. Erhvervsfiskeri i de ferske vande siden middelalderen, pp. 75–86 in E. Hofmeister (ed.) *De ferske vandes kulturhistorie i Danmark.* Silkeborg.

Garner, H. 2002. Klostrene og fiskeriet – især i Gudenåen, pp. 82–93 in H. Goldsmidt and S. Braagaard (eds.)*At leve med de ferske vande – dengang, nu og i fremtiden.* Silkeborg.

Hesel, V. 1993. *Dansk ørrederhverv gennem 100 år.* Skellerup.

Hofmeister, E. 2004. Fiskeproduktion ved de ferske vande – fra karper til regnbueørreder, pp. 75–86 in E. Hofmeister (ed.) *De ferske vandes kulturhistorie i Danmark.* Silkeborg.

Kristensen, H. K. 1973. Ferskvandsfiskeri gennem århundreder. *Sønderjydsk Månedsskrift* 49: 98–114.

Kristensen, H. K. 2004. Klostrenes udnyttelse af de ferske vande, pp. 201–208 in E. Hofmeister (ed.) *De ferske vandes kulturhistorie i Danmark.* Silkeborg: Aqua.

Kristensen, H. K. 2013. *Klostre i det middeladerlige Danmark.* Højbjerg.

Lundberg, S. and Svanberg, I. 2010. Stone loach in Stockholm, Sweden and the royal fish in the seventeenth and eighteenth centuries Sweden. *Archives of natural history* 37: 150–160.

Møller, K. 1953. *Danske ålegårde og andre fiskegårde.* København: Danske Folkemål.

Rasmussen, H. 1959. Fiskedamme og fiskeopdræt, pp. 307–209 in *Kulturhistoriskt lexikon för nordisk medeltid* volume 4. Malmö.

Rockstroh, K. C. 1913. Om ferskvandsfiskeriet i Nordsjælland omkring år 1700. Fra *Frederiksborg Amt* 1913: 1– 43.

Figure 3.1: Crucian carp (Carassius carassius) was one of the farmed fish in the late Middle Ages. (Photo: Lars Nygaard)

Figure 3.2: Skanderborg Castle had fish farming operation in mid-Jutland lakes in the 1600s and there were also fishponds for breeding according to Erik Pontoppidans Den Danske Atlas, 1768.

Figure 3.3: Many fishery managers came from Germany, which had a long tradition of farming carp in ponds. Here is fish master Rudolph Handke, who in 1897 was fishery manager at Gråsten Castle. (Private photo: Nature Agency. Gråsten State Forest)

Figure 3.4: Typical older trout pond. Tågelundgård Fishery. Egtved. 1933. (Photo: C. V. Otterstrøm)

CHAPTER 4

Historical Pond-Breeding of Cyprinids in Sweden and Finland

Madeleine Bonow and Ingvar Svanberg

This chapter describes and analyses the history of pond-breeding of fish in Sweden and Finland (which was an integral part of Sweden until 1809) from late medieval times until around 1900.[1] Very little is known about the history of aquaculture in Sweden and Finland. Most published overviews are superficial. There are very few studies based on sources and hardly anything has been written by historians using modern methods and source criticism. We are therefore uncovering a long, although now broken, tradetion of fish cultivation in ponds which has left scant traces in the written record or the physical environment.

We need to make some clear distinctions about types of aquaculture since much confusion arises from writers not differentiating among natural fish populations in natural or artificial ponds, unselective capture for stocking or storage of wild fish, selective stock and grow operations, and human management of breeding and species-specific stocking and artificial feeding or nutrient management. We deal mainly with the last case. We do not include marine aquaculture, which is a very recent phenomenon in Scandinavia.

The overall purpose of our chapter is to discuss how fish kept in fishponds have been introduced, farmed and spread in Sweden and Finland in early

[1] This chapter was written as part of "The story of Crucian carp (*Carassius carassius*) in the Baltic Sea region: history and a possible future" led by Professor Håkan Olsén at Södertörn University (Sweden) and funded by the Baltic Sea Foundation. We hereby acknowledge him and the other members of the project for their support. We are also grateful to Professor Richard C. Hoffmann (York University), Librarian Leif Lindin (Tierp), Professor Bo Lönnqvist (Jyväskylä University), and Associate Professor Torsten Stjernberg (Zoological Museum, Helsinki) for providing us with important bibliographical information and other helpful advice.

modern times. We want to explore the importance of fishponds and the use of related fish production for food by elucidating their economic, social and religious importance with an emphasis on the historical importance of certain species, in particular crucian carp (*Carassius carassius* (L.)). We have identified several socio-cultural domains during the time period of interest: monastic fishponds in the late medieval times, aristocratic fishponds associated with castles and manors from the late medieval times until at least the early nineteenth century, ponds associated with rectories in the seventeenth and eighteenth century, and urban ponds from the seventeenth century to the nineteenth century. The transformation of the ponds to other functions will also be discussed briefly, as well as attempts to revive cyprinid fishpond production at the turn of the twentieth century.

This chapter aims to illuminate a complex problem that requires a variety of historical methods. We have already stressed that we are dealing with a historical phenomenon that has left few written or physical traces. It is therefore a difficult problem to examine. The study relies on disparate source material. Our source material consists primarily of physical remnants (fishponds), current fish populations which are likely to descend from the former pond production, and onomastic material (such as ichtyonyms and toponyms). One essential source category is the geometrical cadastral maps. Such maps have been shown to be a useful source for studying garden culture, orchards, mills and other economic activities on farms and manors in former times (Nilsson 2010). We have also tried to track down relevant narrative sources such as zoological literature, memoires, historical overviews of manors, provisions lists, old cookbooks and old menus (Svanberg 2010). In Sweden, very few archaeozoological remains from medieval or early modern ponds have been analysed so far (Jonsson 1984, Bonow and Svanberg 2015; cf. Nordeide and Hufthammer 2009).

Early evidence

From the thirteenth century to 1809, Sweden and Finland were united as one country. It therefore seems sensible to discuss the occurrence of artificial fishponds in both Finland and Sweden in the same chapter. The provinces of Scania, Blekinge, Halland and Bohuslän did not become part of Sweden until the Treaty of Roskilde in 1658. We do not know when artificial fish production in ponds was introduced in the Swedish empire. Primitive keeping of fish in more or less artificial settings, which might be termed a kind of rudimentary proto-aquaculture, has probably always occurred

(Bunting and Little 2005). Until rather recently it was a common practice among the peasantry in Finland and Sweden to keep fish in wells dug primarily to provide drinking water. Many small animals thrived in these wells and in order to keep them clean, eel (*Anguilla anguilla* L.) or Northern pike (*Esox lucidus* L.), known as *kaivohauki* 'well pike' in Finnish, were introduced (Sarmela 1994: 52; Svanberg 2000: 91). In Denmark, farmers used the crucian carp for the same purpose (Brøndegaard 1985: 227). This practice is probably very ancient, but has of course nothing to do with economic production of fish. However, it is an interesting example of primitive fish-keeping.

Keeping and stocking fish in man-made ponds is also an old habit. Fish caught by fishermen could be brought to the pond or put in an enclosed area to keep them alive while waiting for later consumption. Fishponds, and simply, ordinary ponds, are mentioned on several occasions in documents from the thirteenth to fifteenth centuries, but it is not possible from these sources to determine if they were used for aquaculture. It seems more likely that they were used merely for stocking fish. These kinds of ponds are mentioned occasionally in documents kept in *Diplomatarium Suecanum*; for instance, in 1241 in Västra Göinge (Skåne) and in 1436 in Sköllersta parish (Närke) as well as in Vaglö (Östergötland) (Wallin 1962: 231; Wiktorsson 1978: 149). An *eel pond* is actually mentioned at Djuråker in Öja parish, Småland, in 1476, which was of course used for stocking eels (Almquist 1938: 167). Millponds are also mentioned relatively often in Swedish medieval sources. However, we do not know if they were also used for breeding fish. Millponds for keeping fish are mentioned from other parts of Europe though (Bunting and Little 2005: 122).

We have two records of lay landowners constructing fishponds from the Swedish area in the fifteenth century. The Charles Chronicle (*Karlskrönikan*) mentions that the military governor Karl Knutsson (Charles, later king of Sweden) commissioned *kroppa damber* ('crucian carp ponds') at Vyborg Castle, a Swedish-built medieval fortress on the Carelian isthmus, in 1446 (Klemming 1866: 245). This first example is constructed at a non-religious locality (the chronicler actually claims that this was the first time a fishpond was founded in the country). Somewhere in northern Gotland a *rwde dam* ('crucian carp pond') was established around 1485 by Privy Council Ivar Axelsson Tott. According to an account book for 1485–87, he bought the fish from local fishermen on the island (Melfors 1991: 173). These two records are among the few examples from written sources in our area of profane fishponds made for the breeding and rearing of fish.

Monastic pond-culture

There is a widespread but uncorroborated opinion that aquaculture was introduced in Scandinavia with the monastic culture. Using ponds to cultivate fish has, according to some European scholars, been important to the monastery economy, not least to readily provide access to fish during the religious period of 150 days when eating meat was forbidden (Fagan 2007: 130–135). Fish was in all likelihood an important part of the diet in the monasteries. Medieval fish bones of cod and related species prove that marine resources were also a source of protein in abbeys in Sweden and elsewhere (Lepiksaar 1965).

The Cistercian Order (Ordo Cisterciensis) is sometimes said to have spread aquaculture to northern Europe, but no evidence supporting this claim is available. Fish certainly played an important role in their diet, but did they really breed fish in ponds, or were the millponds and other kinds of ponds only used for stocking fish? (Johansson 1964: 64–65, 91). Sources of documentary evidence are rare. The monasteries at Alvastra, Gudhem, Gudsberga, Herrevad, Julita, Nydala, Riseberga, Roma, Solberga, Varnhem and Vreta (the latter founded by the Benedictine Order) belonged to the Cistercian Order in Sweden. There are preserved fishponds at the ruins at Alvastra Abbey that actually still contain crucian carp populations, but we do not know when these ponds were constructed. The contemporary ponds at Alvastra are assumed to have been reconstructed in more recent times (Bonow and Svanberg in press).

Very few reliable sources about monastic fishponds exist from Sweden. Certainly, some traces of fishponds are known from Swedish monasteries. From Vadstena Abbey, which belonged to the Bridgettine Order, *rudho damber* 'crucian carp ponds' obviously used for raising fish are mentioned in 1470 and 1517 (Bernström 1969: 441). Similarly, there is still a partially retained pond at the Franciscan convent (Ordo Fratrum Minorum) in Söderköping. Other Franciscan convents with fishponds are located in Linköping. The fishpond excavated at Nylödöse might originally have been part of the Franciscan convent (1473–c. 1520), or, as has been suggested by historian Rune Ekre, from an earlier abbey belonging to the Dominicans. Fishponds are also mentioned at the monasteries at Askeby Abbey, Skänninge Abbey, Gudhem Abbey and All Saints Abbey in Lund (Bonow and Svanberg 2016).

A review of the Swedish monasteries would certainly provide further evidence. Archaeological finds from these monastic fishponds are also few, so hardly any archaeozoological material has been studied (Lepiksaar 1969). One interesting document is the instruction provided by Bishop Hans Brask in Linköping to the state overseer at the Bishop's House, where he was ordered to have a special supervision of *rudadammom* (the crucian carp pond), from where fish were sold on the market. Traces of this pond can still be seen near the cathedral (Bonow and Svanberg 2016).

Some insight into the late medieval monastic fishponds in Sweden is given by Petrus Magni (Peder Månsson) in his manuscript *Bondakonst*, written in Late Old Swedish in about 1520. His name appears for the first time in 1499, when he was chaplain and also the school principal in Vadstena. He was ordained a monk brother in Vadstena Monastery of the Bridgettine Order. In his manuscript from 1520, Petrus dedicates a whole chapter to pond fish culture. The text is to some extent based on Petrus's own experience and provides rare knowledge of pond-breeding of crucian carp and other cyprinids in Scandinavia in late medieval times (Svanberg and Cios 2014).

Fish taxa kept in the ponds

Which species were raised in early ponds in Sweden and Finland? As has been shown above, the most common fish kept and bred in ponds already during medieval times was the crucian carp (*Carassius carassius* (L.)). It has a pond form (*dammruda* 'pond crucian carp') that was earlier recognized as a distinct subspecies or variety. The pond variety was said to be rare in Finland (Lilljeborg 1891; Malmgren 1863: 39).

The ponds were usually called 'crucian carp ponds' (*ruddammar*) on maps; hence we assume that they were primarily used for raising this fish species. The species was well-suited to the Swedish climate and could survive without oxygen through the long winter months in the frozen ponds. It can be transported over long distances even in very trying conditions. In 1756, Anders Tidström (1978: 67) observed in Gothenburg that they could be transported intact over a whole day by placing them in a tub embedded in sphagnum moss (*Sphagnum*). The crucian carp's capacity to survive for long periods without water makes it unique among Swedish fish taxa.

Coupled with their capacity to also survive over winter, frozen in the ice, this makes the crucian carp population persistent even in relatively small pools of water. The crucian carp seems to have been used as a pond fish mainly in northern and north-eastern Europe. In Late Old Swedish it was known as **kroppa* (1446, still in 1520) for this species is known in a late medieval text (Söderwall 1953: 1253; Granlund 1983: 275). The term *Kråppedam* was used for a 'crucian carp pond' in Råda parish in Västergötland according to a cadastral map from 1653, and it is recorded still in the nineteenth century as a local folk name for crucian carp in the same province (Rietz 1862: 357). It is also known from Småland (Lindner 1867: 89). *Kroppe* (recorded since 1554) has also been used in Denmark (Brøndegaard 1985: 227).

The fish is otherwise known as *ruda* in Swedish, but also by its German name *karussa* in south-western Sweden (Skåne, Blekinge, Halland and Bohuslän) and Gotland (recorded since 1556, see Almquist 1911: 586; Sorbonius 1845/1693: 101; Tidström 1891/1756: 50, Kornhall 1968: 103). This ichtyonym is a modification of Middle Low German *karusse*, which is probably of Baltic origin; akin to Lithuanian *karušis* 'carp' (Hellquist 1939: 658). It is known as *ruutana* and *kouri* in Finnish (Malmgren 1863: 38; cf. Kendla 2000: 185 for an etymological discussion of the latter). The use of Danish and Low German loan words as ichtyonyms may indicate that the fish was actually imported and therefore did not originate from local wild populations.

In his zoological lectures from the 1740s, Linnaeus states that the crucian carp was easy to cultivate in ponds and modest in its needs (Lönnberg 1913: 191). Raising crucian carp in fishponds seems to be a Northern European adaptation to the aquaculture practice that was initially developed in central Europe during medieval times. It is also mentioned in both Poland and Denmark in late medieval times (Hofmeister 2004; Makowiecki 2008: 763).

Another species mentioned in medieval sources (Petrus Magni) as a pond fish is the tench (*Tinca tinca* (L.)), known as *swthara* in Late Old Swedish (Granlund 1983: 275), in Modern Swedish *sutare*, locally sutter (Värmland), and in many places as *lindare* (as *lijnnare* from 1612), in the southern part of Sweden also as *skomakare* (Bernström 1972; Svanberg 2000: 266). Wild populations exist in the southern and eastern parts of Sweden (Lilljeborg 1891). It was also known for its capacity to survive transport over long distances. Some were also of the opinion that its presence in fishponds promoted the well-being of other species like the crucian carp (Rothof 1762: 503; Fischerström 1785: 195, 233; Lönnberg 1913: 191; cf. Brøndegaard 1985: 228). However, it does not seem to have been a popular food fish in Sweden during early modern times. It was actually not until the end of the nineteenth century that it became more popular as a pond fish for consumption (Nordqvist 1922: 588).

There is a common but erroneous viewpoint that fishpond culture was introduced in order to breed common carp (*Cyprinus carpio* L.) in Scandinavia. Some of the historical ponds are therefore nowadays wrongly referred to as 'carp ponds' in tourist brochures and popular history writing. On the contrary, this fish taxon has been relatively scarce as a pond fish in Sweden. The species is originally from the Danubian Basin in south-east Europe and spread to Central Europe for fish farming purposes during late medieval times, but was introduced to the Nordic countries more recently (Hoffmann 1995; Makowiecki 2008). The transition of the species from being an exploited captive to a truly domesticated animal took place in the twelfth century (Benecke 2000: 496; Balon 2004: 4–11). We do not have any persuasive evidence for its presence in Sweden until the seventeenth century and since then it has never been common. In 1555, Magnus quite correctly underscores this point when he states that the common carp are missing from Nordic waters (Book 20:23). The Swedish word *karp* (known in its plural from *carpor* from an imported recipe from around 1500 is a German loan word (Bernström 1963: 308). There is a historical view that carp were introduced to Scania, which at that time belonged to Denmark, by the minister and Steward of the Realm Peder Oxe (1520–1575) in around 1560 (Nilsson 1855: 287; Juhlin-Dannfelt 1925: 426). However, as the zoologist Torsten Gislén has shown, this is not supported by any historical sources (Gislén and Kauri 1959: 251–253). In the 1570s there was an attempt to import carp to Kalmar Castle. Carp ponds (*Carpe dammer*) are explicitly mentioned that year. The king gave an order for workers to dig crucian carp ponds ('*som kunne graffwe Rude och Carpe Dammer*') in the castle grounds. Two years later, Arvid Swan is ordered to import live carp from Germany ('*belangendes [...] någre leffwendes karper han ifrå Tydzlandh [...], förskaffe skall*') (Granlund 1876: 86, 98). Some sources mention carp in Sweden during the next century, for instance in 1660. Carp are also mentioned in the royal ponds in Stockholm in 1683 and 1684, when the carp died because they could not survive the winter (Lundberg and Svanberg 2010: 155). The species has never fully adapted to the Swedish climate and it was a continuing problem that carp kept in ponds died during the winter, as Carl Hallenborg (1913) commented from Scania in the 1750s. It was not until the eighteenth century that there is convincing evidence that they are cultivated on some large estates in southern Sweden (Linnaeus 1751: 224; Rothof 1762: 232). However, as Fischerström (1761) observes, common carp are still rare in Halland manor ponds. Towards the end of the century (the 1790s), however, there are reports that they had been introduced at Dömestorp Manor and Vallen Castle in Våxtorp, Halland (Osbeck 1996: 64). Contrary to Fischerström's (1761) assertion, Osbeck (1996: 64) claims that 180 common carp were introduced at Våxtorp in the 1680s or even earlier. However, there is no source that confirms that the introduced carp survived and were cultivated in any ongoing way. In the eighteenth century, most carp consumed in Stockholm were imported from Danzig (Lönnberg 1913: 191). In Finland carp were not introduced until the 1860s (Piironen 1994: 70).

On royal estates, and possibly some manors owned by the aristocracy, attempts were made to breed the small but tasty stone loach (*Barbatula barbatula*). It seems to have been introduced as early as the 1680s. Its old Swedish name *smerling*, which is used in the historical sources, shows its German origin. The records indicate that stone loach were kept in ponds in The Royal Game Park (Kungl. Djurgården) in Stockholm at that time. They were first kept in a separate pond but later relocated to another pond and cohabited with carp. This is confirmed by an annotation from 1683, which indicates that three men were paid for two days' work to relocate stone loach from their pond to the carp pond in The Royal Game Park. Stone loach had a reputation for being easy to digest and suitable for sensitive stomachs. This species of fish was therefore a popular dish among the royals. Sixty years later, in 1740, King Frederick I is said to have released stone loach in ponds at Ulriksdal Royal Castle. The fish were possibly imported from the king's native Germany to be eaten as a delicacy (Lundberg and Svanberg 2010). The cultivation of stone loach is also mentioned in relation to the royal fishponds in Denmark in the mid-seventeenth century (Hofmeister 2005: 76).

Some fish taxa may have also been kept by the royals and the social elite for non-consumptive purposes such as ornamentation. For instance, sterlet (*Acipenser ruthenus* L.) and European weather loach (*Misgurnus fossilis* (L.)) were kept in ponds in the Royal Gardens in the 1740s for this purpose (Lundberg and Svanberg 2016). There were even attempts to introduce them in Lake Mälaren (Bernström 1947). Keeping of ornamental fish, such as the gold varieties of the cyprinids known as goldfish (*Carassius auratus* (L.)), golden orfe (*Leuciscus idus* (L.)) and golden tench, in ponds to enhance garden aesthetics is a more recent trend. Ornamental fish of these kinds are not mentioned in Swedish sources until the eighteenth and nineteenth centuries (Lilljeborg 1891: 158–160; Svanberg 2007: 76–77).

Historically, cyprinids have been kept in ponds for economic reasons, but never with any success for aquaculture. There are occasional recorded sources of keeping roach (*Rutilus rutilus* (L.)), asp (*Aspius aspius* (L.)) and other species like perch (*Perca fluviatilis* L.) and pike (*Esox lucius* L.) in ponds. In all likelihood these refer to the stocking of these species rather than breeding them in the ponds. For instance, a special pond established for asp in Uppsala is mentioned in 1590 (Bernström 1969: 442). We should also remember that Schroderus (1640: 167) refers to *rudammsfiskar* ('crucian carp pondfish'), a label he uses collectively and therefore imprecisely to refer to carp, ide, asp, pike and crucian carp. These were probably fish taxa that could be stocked in a crucian carp pond.

After the reformation

There is a widespread myth that the Protestant Reformation in the 1520s and 1530s was in itself a threat to fish farming in Sweden. However, we cannot see any evidence to support this view of a decline in aquaculture in early modern times. On the contrary, the evidence indicates that pond culture not only survived but was also extended to other environments during the sixteenth and seventeenth centuries. This, however, is a record from 1548 of the existence of a crucian carp pond in a monastery in Vadstena. The oldest written information we have about fishponds from Alvastra also dates back to 1548. This is after the Reformation and the source indicates that the Royal Court paid 25 *marker* in salary to soldiers to maintain the crucian carp ponds there, which indicates that Gustav Vasa supported fish farming in the country. Three ponds can be seen maps from 1640 and 1691 of the former site of Alvastra Abbey. On a map of Askeby from 1776, at least one fishpond is visible. It is not clear whether the fishponds discussed above were used for raising fish and it is also not known if they are the same as the ones that were used before the abbeys were abandoned.

Pond cultivation in castle and monastery grounds that had been established in Sweden during late medieval times lived on as a form of production after the Reformation and even spread to other socio-cultural domains. Scattered data in the sources shows that several castles had crucian carp ponds in the sixteenth century. A crucian carp pond at the castle in Turku is mentioned in 1552 (Gardberg 1959: 76). When King Gustav Vasa visited Örebro in August 1554, the castle was repaired, a new kitchen built and a crucian carp pond cleared (Nordström and Dahlander 1913: 31). His son John III of Sweden tried with foreign help to construct carp ponds at Kalmar Castle in the 1570s (Silfverstolpe 1876:86). In 1570, a crucian carp pond was constructed at the Royal Castle in Uppsala (Bonow and Svanberg 2012: 134, 140). Fish production in ponds seems to have been commonplace in the grounds of Swedish castles during the sixteenth and seventeenth centuries (Nilsson 1939).

In the early 1690s, Åke Claesson Rålamb published his encyclopaedic *Adelig Öfvning*, which is a kind of handbook on agronomy, among other topics. The intended readership of the handbook was young noblemen. In his book, he also describes the construction of crucian carp dams, stressing for instance the fact that the pond must have a breathing hole in winter and that it should be maintained with horse manure (Rålamb 1691: 96). Isaacus Erici (1576–1650), a priest in Stenby parish in Östergötland, translated a German handbook on gardening and household economy, which promoted the expansion of crucian carp ponds (Erici 1683: 169). In his magnificent baroque epos, *Guds Werk och Hwila* ('God's work and rest') from 1685, the Bishop of Linköping, Haqvin Spegel, advocates fishponds with carp and crucian carp (Spegel 1998).

There are numerous data from the early seventeenth century on crucian carp ponds in castle grounds (Ellenius 1967: 69). The accounts from Gripsholm Castle in Mariefred mention that in 1620 two crucian carp fishponds existed in the large castle garden (Bonow and Svanberg 2011). A crucian carp pond was built on the Skälby Estate in Kalmar County in the 1640s (Hofrén 1937: 120–121). Svartsjö Palace on the island of Färingsö in Lake Mälaren still has a preserved crucian carp pond that probably dates back to the seventeenth century. There is an unverified story that a crucian carp pond was constructed on the roof of Bogesund Castle in Uppland in the 1640s (Nisser 1927: 31). A 1674 map of Skokloster Castle located on Lake Mälaren shows a crucian carp pond beside the old hop-garden (Ellenius 1967: 70).

At Visingsborg Castle on Visingsö, an island in Lake Vättern, Count Per Brahe the Younger (1602–1680) constructed crucian carp ponds, which can be seen on cadastral maps. Fishponds are also reported from several locations in Finland, including Jacob de la Gardie's estate in the town of Nykarleby where his wife Ebba Brahe raised crucian carp (Huldén 1957). Astronomer Tycho Brahe (1546–1601) also had a large number of ponds on the island of Ven in the late sixteenth century. The island was still under Danish rule at that time (Nilsson 1939).

Fishponds in manorial culture

Interest in pond fish continued into the Age of Liberty (1718–1772) in Sweden, as many reports from the provinces show (e.g. Barchaeus 1924: 60; Fischerström 1761: 266; Fischerström 1768: 195; Osbeck 1996: 65). There appear to have been fishponds at numerous manors in southern Sweden and in Finland during the eighteenth century (Schwerin 1932: 157). Lindgren (1939: 60) suggests that watermill dams might have also functioned as fishponds. There is nothing, however, in the sources that support this opinion.

In Scania we can observe an interest in large-scale fish farming in the eighteenth century on some of the larger estates. In Skånska resa (1751), Carl Linnaeus described the estate of Marsvinsholm in Scania where there were 99 ponds, including one on a roof, containing common carp and crucian carp. What significance crucian carp ponds had for supplying estate households has not been investigated but these systems of ponds were obviously built in order to sell fish. During his tour of Scania in 1749, Linnaeus noted several estates, including Marsvinsholm, Vrams Gunnarstorp and Lärkesholm, where there were fishponds that grew crucian carp and common carp. He also provides us with some production figures (Linnaeus 1751: 224, 256, 378–381).

The Trolleholm Castle in Scania is interesting in this context because surviving local archival documentation provides detailed data on crucian carp production. During the seventeenth century several fishponds were constructed, both near the main building and distributed throughout the grounds of the estate. In addition to the carp production records, a diary was maintained with details of the estimated number of fish fry released into the ponds. These data gives us insights into the magnitude of production in the ponds. In 1799 there were about twenty fishponds on Trolleholm. In the Albergsdammen, 270 crucian carp were caught on 28 June and 2 July 1806. On 24 August 1808, 236 crucian carp fry were released and on 28 August 1815, 100 larger crucian carp (and a number of pike) were caught. The Albergsdammen was stocked with crucian carp on 8 September 1800. On 10 October 1804, no fewer than 160 crucian carp and three tench were caught and on 17 July 1809 twenty crucian carp and three pike were caught. On 23 July 1810 the catch was 48 crucian carp, one large tench and several small fish. We also know that Trolleholm had special wardens for the management of fishponds. Fishing in the ponds at Trolleholm is recorded for the last time in 1817 (Bonde-Trolle 1905: 146–147).

Rectory fishponds

One important discovery in our investigation is the presence of crucian carp stocked fishponds at the rectories. This has been observed before, but never investigated in detail (Arvastsson 1977: 41). It was obviously not just rich landowners who raised crucian carp in ponds for household consumption. Why this clerical aquaculture developed is not clear, but it has nothing to do with religious fasting, a custom that was abandoned after the Reformation (Baelter 1783: 209–211). The fishponds were probably part of the rectory economy, of which little is known. It seems to have been a widespread pattern in some parts of southern and eastern Sweden (Bonow and Svanberg 2014).

While source material is a little ambiguous on this point, clerical pond cultivation of crucian carp was established in the seventeenth century. Cadastral maps from 1640 show virtually no ponds in parsonages, which may indicate that pond farming had not yet been established, but the lack of ponds could also be attributed to survey issues. There are only a few instances of fishponds appearing on the early cadastral maps, for example, in 1641 as part of the Stora Tuna Rectory in Dalecarlia (Bonow and Svanberg 2011).

The construction of ponds for fish farming in vicarages is infrequently mentioned in the sources. Peder Berger, a vicar in Runtuna Parish in Södermanland, hired two Dalecarlian men in 1662 to dig a crucian carp pond. In 1679, at Dunkers Vicarage in the same province, a crucian carp pond was constructed in the garden (Flinck 1996: 123).

We know that there were crucian carp ponds in vicarages in several provinces in the Swedish countryside from the late seventeenth through to the eighteenth centuries, but preserved maps and surveying documents distinguish two important clusters in Scania and Östergötland. We have also found evidence of fishponds on the cadastral maps in rectories in Uppland and Västergötland as well as in Småland, Närke, Dalecarlia and Gotland (e.g. Vall Parish, which still has a pond stocked with crucian carp). It is possible that further research will provide a more nuanced view, but here we must content ourselves with a brief presentation of the clerical crucian carp ponds in Scania and Östergötland (Bonow and Svanberg 2011, 2012: 137–139).

There is rich data available on Östergötland in the eastern part of Sweden. One important source is the cadastral maps from the late seventeenth and the first half of the eighteenth centuries. They give us information about pond locations. Approximately twenty vicarages had crucian carp ponds during this period. We can distinguish two areas where crucian carp ponds seem to have been common. These are around the great lakes, Vättern and Roxen, and the neighbouring towns of Söderköping, Norrköping and Linköping, where priests had constructed fishponds either in their gardens or in adjacent areas (Bonow and Svanberg 2011).

Parsonage garden culture emerged in the late seventeenth century and Söderköping's rectory is considered to be one of the first in the area to have a large garden. It was recommended that priests establish kitchen gardens with vegetables and herbs, orchards and if water was available, construct crucian carp ponds (Cnattingius 1932). Östergötland archival sources show that crucian carp ponds had ceased to be constructed by 1750. They slowly disappear from the maps from 1800 onwards. Today only a few of these ponds remain, some of which contain crucian carp, for example, Styrstad Parish (Bonow and Svanberg 2011).

In Scania we find many crucian carp ponds in rectory grounds in the late seventeenth century (Arvastsson 1977: 41). Olof Bertelsson Aquilonius (1630–1684), a vicar of Löderup Parish, shows that clerical fish farming could have quite significant economic importance. He was probably more of an entrepreneur than a spiritual adviser. He had a private boat with which he transported crucian and other carp from his ponds to sell in Copenhagen (Cavallin 1857: 82).

A survey of Gladsax Parish in 1699 shows crucian carp ponds located on the outfields were already defunct at the time of the mapping. The six ponds depicted on the 1699 map had all belonged to the Royal Palace. Within the village, the rectory still had two ponds in use (Nilsson 1939). A crucian carp pond was still operational there in the 1756, when Anders Tidström passed through the village. He stated that a problem with ponds on the outfields was that someone could steal the fish, but also that waterfowl could bring fry of predatory fish (pike) that would threaten the fish stock. In the middle of the village, fishponds could however give an output and could also be emptied to take advantage of the manure that was allowed to drain into the ponds (Tidström 1891: 50). The 1693 Veberöd cadastral map shows that 10 of the twenty-five homesteads in the village had crucian carp ponds on their property. Unlike other parts of Sweden, where ponds disappeared during the eighteenth century, they persisted in Scania and some even became the subject of litigations in the land reforms of the nineteenth century (Nilsson 1939).

Urban fishponds

A very interesting finding in our search for fishponds in Sweden is the occur-rence of crucian carp ponds in or near cities during the eighteenth and nine-teenth centuries. Urban traces of carp ponds are found in city areas, for example, in Stockholm and Uppsala, and in street names, for example, in Stockholm, Eskilstuna, Gävle, Mariefred and Lindesberg, throughout Sweden. In Gävle, for example, there is Ruddammsgatan, which is a street located in the area where a crucian carp pond was located at the beginning of the eighteenth century. Crucian carp ponds and other fishponds are also mentioned in historical sources from Eskilstuna, Arboga, Örebro, Uppsala, Norrköping, Linköping, Varberg and Ronneby. Many of these fishponds are shown on cadastral maps. Abraham Hülphers (1783:83–84) refers to the pond in

Eskilstuna as a 'crucian carp pond' in 1783, and it was still inhabited with fish in the 1920s. Other urban crucian carp ponds also existed at Almrothska ängen and at Gästis in Eskilstuna, but they were gone by 1920. Traces of an urban crucian carp pond in Arboga were still apparent in the late nineteenth century (Bonow and Svanberg 2015).

Ruddammen is a well-known area in Östermalm in Stockholm. At the beginning of the 1700s there were several fishponds here and the largest of them belonged to the inn-owner Ingemar Frodholm on his property Inger-marshov. Women also owned crucian carp ponds. From the early seventeenth century there is evidence that a widow named Elsa Hoffman owned property with a crucian carp pond on the outskirts of Örebro (Lenander Fällström 1987: 111). Other instances of urban crucian carp ponds included Marieberg on Kungsholmen (Wikström 1840: 14) and Uppsala, where there were several crucian carp ponds, one of which is still remembered through Rudan, the name of the city block in the centre of Uppsala (Bonow and Svanberg 2012: 141). Carl Linnaeus also mentions the existence of crucian carp ponds and fishponds in Uppsala. For instance, several of the ponds in Uppsala were owned by a J.D. Fick (Linnaeus 1755; Triewald 1746). The old royal fishpond in Uppsala was an urban pond in the eighteenth century (Linnaeus 1899: 36).

According to the zoologist Sven Nilsson (1855: 295), there were still numerous crucian carp ponds in Lund up until the 1850s. In the 1760s, a complex of fishponds for breeding carp was built on Helgonabacken in Lund. Remains of one of these ponds are still discernible in the park outside the university library. The extent of crucian carp ponds in towns, their ownership and management still remains to be explored.

A special case of urban ponds were those created for the purposes of breeding medicinal leeches (*Hirudo medicinalis* L.). There are several instances of such ponds (Malm 1863: 175). The street name *Igeldammsgatan* in Kungsholmen in central Stockholm reminds us of such ponds. In 1835, no fewer than 40,000 leeches were introduced in the ponds, which were run by Apotekarsocieten in Stockholm. Leeches were used in large numbers in eighteenth and nineteenth century medicine (800,000 in Sweden and 200,000 in Finland in 1850) and to satisfy this high demand they were farmed in ponds (Linnaeus 1764; Whitaker et al. 2004). Ponds for breeding leeches still existed on Hisingen, near Gothenburg, in the early twentieth century (Ahlbäck 2006).

Construction and management of ponds

The oldest Scandinavian fishponds at monasteries are said to have been constructed with the continental carp ponds as models (Rasmussen 1959). Very little is known about the construction of these fishponds in Sweden. A pond was excavated at Lödöse Convent in 1964, but no analyses have yet been published (Ekre 2007). This is also true of fishponds in other socio-cultural domains. We can still see some of them at a number of manors and rectories, but they have been changed several times since they were actually used for aquaculture. Some handbooks give detailed information on how to construct ponds. Probably the most detailed was published by Carl Henrik König, with a chapter on various kinds of ponds (König 1757: 126–131). Schultze (1778: 205–224) also gives many interesting details on how to construct ponds.

More interesting is the material edited by Carl Knutberg in 1768, based on descriptions sent to the Swedish Academy of Sciences. He differentiates between ponds for the breeding and rearing of fish, which he named *plantér-dammar* ('rearing ponds'), and ponds for just holding captured fish for later consumption, which he called *sump-dammar* ('nurse-ponds') (Knutberg 1768). A commentator at the time, Schultze (1778: 217) was of the view that all kinds of fish could be kept in the latter kind.

There were obviously two kinds of ponds for breeding fish. The first type was located in orchards. Several of these ponds have survived, although they are now used for other purposes. It was more common, however, at least at the manors, to dam a brook or a small stream. These kinds of ponds have not survived until today. Knutberg (1768) gives many details about the construction of ponds. Some other sources also provide details about the construction of the ponds. A document from 1658 describes how a beam (*rud-dammsbalk*) was used as a partition in the ponds (Hultman 1913: 236). As far as we know, no technical or archaeological studies have, been made of those old ponds that still endure in Sweden.

A cadastral map from Höja Manor in Uppland shows that trees were planted around the pond in order to give shade (Ulväng 2009: 80). Such planted trees can also be seen on several cadastral maps from elsewhere. The planting of willow or other broad-leaved trees to give shade to fishponds is also recommended by Knutberg. The trees also helped to protect the fish from birds of prey (Knutberg 1768: 174).

Details regarding pond cultivation are sparse. The growth of the crucian carp in ponds had already been discussed by Olof Rudbeckius in the seventeenth century (Rudbeck 1947: 252). Knutberg (1768) discusses the construction of ponds. Tiburtz Tiburtius (1706–1787), a priest from Vreta Parish, conducted extensive experiments with fish farming in his parish. Furthermore, he discussed his experience in an article published by the Swedish Academy of Sciences in 1768. He owned several custom-dug ponds stocked with crucian carp and tench, but he was apparently dissatisfied with the production outcome and so instead sought to develop a fish farm system of natural lakes (Tiburtius 1768).

The large fish farm at Marsvinsholm Castle in Skåne is described in some detail in Linnaeus's 1751 travelogue. At the time of Linnaeus's visit there were almost forty ponds at Marsvinsholm, which were all stocked with common carp and crucian carp. However, a few years earlier there were as many as 99 ponds in the grounds, as well as a pond on the roof, which was made of lead. Further details about the construction and management of fishponds in Lärkesholm are also described by Linnaeus. These two fish farms were the largest in Sweden at the time of Linnaeus' visit in1751 (Linnaeus 1751: 254, 370–380).

Handbooks also contain information regarding crucian carp ponds. Lorens Wolter Rothof stresses that they should be kept in nutrient-rich ponds and that the ponds be landscaped to allow the manure to flow into them. He suggested that under these conditions the fish would grow large and multiply quickly. According to Rothof (1762: 391), each square fathom (= 3.17 m^2) could yield a barrel of crucian carp per month. In less nutrient-rich water, however, the fish reproduced slowly and were only a few inches long (Rothof 1762: 391). The use of manure-laced waste water seems to have been an important part of fishpond culture in Sweden (Tidström 1891/1756: 50; Fischerström 1785: 196; Barchaeus 1924: 60).

Johan Fischerström, a prolific economic writer, gave an interesting insight into crucian carp breeding in ponds in 1761 with the following advice on how to tend them: "They love clay and grass floor. Newly dug ponds ought either to be clothed with weed or sown with oats. There should be a mother-pond, and a couple of other ponds for males only. It is customary to throw balls, made from mash and blue clay, into the ponds. Thick sour milk makes them particularly fatty and delicious" (Fischerström 1761: 266).

This quote shows that they had detailed knowledge of all the different aspects of cultivation in crucian carp ponds. Another piece of advice was to loosen "the scale on either side, then cut out a piece of the tail, which promotes their growth" (Fischerström 1761: 266). At the time there were apparently strong views on how the ponds should be constructed, about feeding regimes and how the fish could be manipulated to stimulate growth.

Carl Ulrik Ekström (1831: 199), a vicar in Södermanland in the early 1830s, described how crucian carp were harvested from ponds with the help of hand nets or fish traps baited with eggshells. On one occasion he had observed a specially-made dragnet for harvesting crucian carp from ponds. In the Scandian ponds, the crucian carp was caught with a dragnet or a special wicker basket drawn along the bottom of the pond (Nilsson 1855: 296). A seventeenth century document from Finland mentions a fish trap for capturing crucian carp in ponds (Hultman 1913: 236).

According to eighteenth century authors, the presence of smooth newts (*Lissotriton vulgaris* (L.)) and leeches (*Piscicola geometra* (L.)) was a problem in fishponds. Both leeches and newts were accused by Mårten Triewald, Carl Linnaeus and Johan Fischerström, among others, of causing considerable damage to the crucian carp stock in fishponds. Linnaeus suggested adding some salt to the water to drive the newts away, something that Triewald, who had a thriving crucian carp pond at Elisabethsberg on Kungsholmen in Stockholm, confirmed as a successful strategy during his own experiments (Triewald 1746; Cederlöf 1766: 17; Fischerström 1785: 236).

Otters were also seen to be a problem for fishpond owners, and manors kept dogs to keep them away (Knutberg 1768: 178). Beavers and water voles are also mentioned as pond pests (Schultze 1778: 215). Pikes that were accidentally introduced into the fishponds were also dangerous (Linnaeus 1751: 379).

Farmed fish for food

The Swedish Royal Kitchen and the high nobility were also fond of crucian carp for the table. We have a few scattered reports from the Royal Kitchen in the 16th century which mention various dishes like 'fried crucian carp with apples' (*Steckta karusser och äplar* 1556), 'crucian carp pâté' (*Charutze Pasteijer* 1600) and others. Dishes of carp and milt of carp are also mentioned, including carp-tongue, which is actually a fat-like formation in the throat of the fish, which was used for various delicacies in the early seventeenth century such as carp-tongue pâté (*Pastei af Karpetungor*) and carp soup (c.f. Rålamb 1690: 120, 107; Anonymous 1730: 21).

Crucian carp were served at manors and rectories in the eighteenth and nineteenth centuries. It was easy to harvest from a well-managed pond and the carp provided a good meal (Reuterholm 1909: 137; Lönnqvist 1993: 20–21; Roberg 1951: 201). At Möllershof manor in Mäntsälä in Finland there is clear evidence that crucian carp were served in connection with the Möllersvärd family's funerals. Among the many dishes recorded as being served, we also find that crucian carp were served as part of the very elaborate and ritualized funeral dinner gourmet dishes (Hausen 1915: 184). In 1653, at a wedding in a bishop's household in Jutland in Denmark, no fewer than 2,380 crucian carp were served as part of the celebrations (Möller 1871: 8b).

In the cookbooks from the seventeenth and eighteenth century, we find considerable evidence of crucian carp as a valued part of the culinary culture. In her *Hjelpreda i Hushållningen För Unga Fruentimber* Kajsa Warg, probably Sweden's most famous cookbook author, gives several recipes containing crucian carp (Warg 1755: 292). Other popular cookbooks also have various carp recipes (e.g. Björklund 1808: 109 and Hollberg 1896: 129). However, we do not know anything about how crucian carp were utilized within the rectories' kitchens.

In his fishing guide of 1778, Samuel Th. Schultze described the flesh of carp as "nice and tasty" (Schulze 1778: 79). The crucian carp's popularity as a fish for the table lasted well into the nineteenth century. Ichthyologist and Vicar C. U. Ekström praises its flavour and consistency (Ekström 1831: 199). As late as 1855 the zoologist Sven Nilsson claimed that, "It is considered a very tasty fish" (p. 293), and describes how it was boiled and served with white cream sauce flavoured with horseradish (Nilsson 1855: 293). Some authors also commend the flesh of the tench. It was said to be white, juicy, satisfying and easy to digest (Bergius 1787: 314; Fischerström 1785: 195).

The end of an era

Carl Linnaeus, always alert to ideas that could benefit the country's economy, was obviously enthusiastic about the future of aquaculture in Sweden. In his journey to Scania, published in 1751, he was clearly inspired by the carp and crucian carp cultivation he had studied in Lärkesholm. He was so impressed that he thought that carp cultivation could also have a future in other parts of Sweden. His thoughts about this are clearly expressed in the following statement urging remedial action on this matter: "Our Nation should think again about this matter, which so far at least up in the country, has not engendered the respect it deserves, then so wonderful opportunities for the fish once could be enough valued to its satisfaction, and the mountainous landscape, which gives smaller grains, could replace the loss with fish" (Linnaeus 1751: 379).

Linnaeus's observations in Scania led to an increase in interest in aquaculture among authorities and economic writers. A doctoral thesis on Scanian carp ponds, under the presidium of Claes Bleckert Trozelius, was defended by Olof Cederlöf at Lund University in 1766. Although based mainly on Linnaeus's writings it gives some insights into how aquaculture was perceived by economists at the time (Cederlöf 1766). The same year, the Swedish Academy of Sciences described the best way of constructing ponds for fish farming (Lindroth 1967). Four manuscripts on the issue were prepared, edited and published by the industrious Carl Knutberg in 1768 (Knutberg 1768). Also a royal decree, issued in November 14, 1766, requested Sweden's governors to encourage the construction of fishponds for the cultivation of asp, carp, crucian carp and ide (Kungl. Maj 1766). Agronomic economy writers published descriptions of how to construct fishponds and raise common carp and crucian carp with texts based mainly on Linnaeus's travelogue from 1751 (Carleson 1768). As early as 1760 a Professorship of Practical Economy (Borgström Professorship) was established at Uppsala University with the task of lecturing not only on gardening and hunting but also on fishery and aquaculture (Lindroth 1975).

However, all these efforts were in vain. The economic margins of aquaculture were too small, so fish farms never fulfilled their great promise. Linnaeus's enthusiasm notwithstanding, the cultivation of crucian carp, common carp, tench and other fish that occurred in the grounds of castles, manors and rectories and in some cities had already faded by the early nineteenth century. This is related to the Agrarian Revolution where more and more land was cultivated to increase agricultural productivity. Fish prices were also very low during this period. As a result, it was deemed not to be worth the effort to cultivate fish for food (Gadd 2011; Nordqvist 1922: 590).

During this time remnant ponds were filled in (which is sometimes indicated on the cadastral maps with names like Ruddamsängen 'the crucian carp pond meadow') or were turned into ornamental ponds, especially when the English landscape parks became fashionable in Sweden after around the 1780s. Sometimes the old fishponds would be used to keep ducks or, in the grounds of the larger mansions, swans. The mute swan (*Cygnus olor*) was distributed as an ornamental bird on estates during the second half of the eighteenth century and became an important element of park landscapes (Flinck 1996: 76, 86; Svanberg 2007: 86–87).

A renewed interest in aquaculture

We can discern a renewed interest in aquaculture n the mid-1800s when Baron Gustaf C. Cederström was commissioned by the Royal Academy of Agriculture in 1856 to travel around the country taking stock of aquaculture for fish production in the Swedish countryside. On his journeys, he found small initiatives here and there. Of particular interest to us is that he describes the presence of isolated, old, overgrown fishponds that had recently been restored at James Steffenburg's property, Lövnäs, near Falun and at a castle near Tidaholm. Experiments with new pond fish species such as asp, perch, bream and ide occurred among some enterprising households at mansions and on estates but he did not encounter any functioning crucian carp ponds. The cultivation of carp for subsistence purposes, which had been present at the seventeenth and eighteenth century rectories and manor houses, was however gone (Cederström 1857: 13).

Influenced by French success with trout breeding, in 1858 Cederström initiated an aquacultural experiment together with John Lenning (1819–1879) at Holmen near Norrköping. At about the same time, Carl Byström introduced aquaculture in Jämtland (BiSOS 1861: 20). These and subsequent forays were focused on farming salmonids (Schött 1914: 394–395). Handbooks in aquaculture were also published (Norbäck 1884; Trybom 1885).

The late nineteenth century saw a commercialised resurgence of the cyprinid fishpond culture. In 1879, Carl Wendt, a landlord, established extensive carp farming on Gustafsborg Estate in Perstorp in Kristianstad County. In all he constructed 63 ponds over 356 acres. Wendt was originally from Germany and his ideas were probably derived from his experience there. Some of his neighbours were also inspired by his fish farm initiative and founded their own ponds. His son Wilhelm Wendt moved to Lammhult in Småland where he founded a fish farm stocked with common carp and tench (Trybom 1885; Nordquist 1922: 591). In 1890, the state authorities founded a fish farm called Fiskodlings- och sötvattensbiologisk anstalt at Finspång in Östergötland headed by a biologist and ichthyologist named Rudolf Lundberg (1844–1902) (Anonymous 1892). MP Carl M. Peterson provides a vivid description of his rotation strategy when cultivating crucian carp and tench in Småland in the early twentieth century (Edling 1910: 22).

Under the leadership of fisheries commissioner Oscar Nordqvist, Södra Sveriges fiskeriförening founded an experimental and model farm for aquaculture in Aneboda (Småland) in 1906. These ponds, managed by state authorities, were primarily stocked with carp and tench. These ponds are still operational today and they produce carp for a variety of commercial purposes, including common carp for recreational fishing, ornamental Koi for garden ponds, and grass carp (*Ctenopharyngodon idella* (Valenciennes)), for weed and algae control in ponds, for example at Kolmården Zoo. With state support, educational programmes in fish farming were held in schools at Ängelsberg in Västmanland, where fish farming of rainbow trout (*Oncorhynchus mykiss* (Walbaum)) was first practised in Sweden, and in Aneboda (Ahlbäck 2006). The Swedish Rural Economy and Agricultural Societies (*hushållningssällskapen*) participated in these educational efforts (Larsson 1922; Alm 1927).

In total, in 1916, there were fish-farms on almost 1,700 hectares in the counties of Kristianstad, Kronoberg, Malmöhus and Halland in southern Sweden. Two thirds of the area used for fish farming was located in Kristianstad County. Fish farms were also beginning to be established else-where in the country. These were mostly salmonid fish farms in Örebro and Jämtland counties. However, there were also quite extensive farms in Gimo-Österby, Uppland (Nordquist 1922: 591). A fish farm focusing on raising common carp, crucian carp and tench was founded by J. Albert Ahlbäck in 1917 at Svankällan on Hisingen outside Gothenburg. He also published a booklet on tench in aquaculture (Ahlbäck 1931). The fish farm on Hisingen was operational until 1974.

The focus of aquaculture in the early twentieth century was on salmonids, common carp and tench (Lindstedt 1912; Nordquist 1922). The small crucian carp in the ponds were no longer of interest for food and their only economic significance was as bait and perhaps less so as aquarium fish. It was common for young boys in their games to move crucian carp from ponds to other small bodies of water and it was by this means that they were introduced into many small lakes all over Sweden and Finland (Sundman 1989: 6; Andersson 1942: 441). The impact of pond crucian carp on the wild populations is not known (cf. discussion in Moyle 1997). Tench have also been introduced into small lakes across Sweden and Finland (Lilljeborg 1891: 174).

Although cyprinids continued to be caught by fishermen and eaten during the first decades of the twentieth century (Trybom 1895), consumers gradually began to prefer sea fish on the table. At about this time, railway transportation made marine species (cod, haddock, whiting and flatfish) easily available and inexpensive all over the country, making fish from lakes and cultivated from aquaculture less popular. At the same time, there was a decline in in interest in cyprinids as food in Sweden. Although some wild species (bream) were still eaten, they were general rejected by modern consumers in after the end of World War II and they disappeared from the dining table. Only small ethnic enclaves continued to demand and consume cyprinids (Ståhlberg and Svanberg 2011).

Final remarks

The crucian carp had its heyday as a food fish in Sweden and Finland from the Late Middle Ages until the late eighteenth century. It has never regained the status that it enjoyed during this period despite several efforts to increase its popularity. The decline of crucian carp as a prominent table fish is mirrored throughout Western Europe. There is some patchy evidence from Finland that indicates that crucian carp taken from the wild are still occasionally used for the table. For instance, Professor Bo Lönnqvist, in correspondence from 3 March 2011, assured us that he still fishes for crucian carp in Strömfors in Eastern Uusimaa in Finland. The correspondence further describes how he prepares them for the table by smoking them. According to Lönnqvist this should be done before Midsummer (Lönnqvist *in litt.* 2011).

Remnant fishponds with stocks of crucian carp at manors and rectories are part of the biocultural heritage in Sweden and Finland and deserve to be preserved.

References

Ahlbäck, H. 2006. *100 år 1906–2006 i Aneboda*. Lund.

Ahlbäck, J. A. 1931. *Sutareodling jämte allmänna anvisningar om insjöfiske*. Göteborg.

Alm, G. 1927. *Fiskodlingens ordnande inom landet: föredrag vid Föreningens Sveriges fiskeriinstruktörer och tillsyningsmän årsmöte i Stockholm den 21 mars 1927.* Stockholm.

Almquist, J. A. 1911. *Konung Gustaf den förstes registratur Vol. 26. 1556.* Stockholm.

Almquist, J. A. 1938. *Arvid Trolles jordebok 1498.* Stockholm.

Andersson, K. A. 1942. *Fiskar och fiske i Norden* vol. 2. *Fiskar och fiske i sjöar och floder.* Stockholm.

Anonymous 1730. *Oeconomia eller hushålds-book, sampteliga thet ährbara qwinnokiönet til hielp.* Stockholm.

Anonymous 1892. Fiskodlings- och sötvattensbiologiska anstalten vid Finspong. *Svensk Fiskeri-Tidskrift* 1: 1–6.

Arvastsson, G. 1977. *Skånska prästgårdar.* Lund.

Baelter, S. 1783. *Historiska Anmärkningar om Kyrko-Ceremonierna, Så wäl Wid den offentliga Gudstjensten, Som Andra tilfällen hos de första Christna, och i Swea Rike, I synnerhet Efter Reformationen til närwarade tid.* Stockholm.

Balon, E. 2004. About the oldest domesticates of fish. *Journal of fish biology* 65 (Supplement A): 1–27.

Barchaeus, A. G. 1924. *Underrättelser angående landthushållningen i Halland.* Lund.

Bennecke, N. 2000. *Der Mensch und seine Haustiere. Die Geschichte einer jahrtausendealten Beziehung.* Stuttgart: Thesis.

Bergius, B. 1787. *Tal om läckerheter, både i sig sjelfva sådana, och för sådana ansedda genom folkslags bruk och inbillning* vol. 2. Stockholm.

Bergström, G. 1892. *Arboga krönika* vol 1. Örebro.

Bernström, J. 1947. Om försöken på 1730-talet att i Sverige inplantera sterlett. *Fauna och flora* 42: 145–148.

Bernström, J. 1948. Bidrag till kännedom om några svenska fiskar i äldre tid. *Fauna och flora* 42: 35–148

Bernström, J. 1963. Karp. Sp. 307–308 in *Kulturhistoriskt lexikon för nordisk medeltid* vol. 8. Malmö.

Bernström, J. 1969. Ruda. Sp. 440–442 in *Kulturhistoriskt lexikon för nordisk medeltid* vol. 14. Malmö.

Bernström, J. 1972. Sutare. Sp. 444–445 in *Kulturhistoriskt lexikon för nordisk medeltid* vol. 17. Malmö.

BiSOS 1861 = *Bidrag till Sveriges Officiella Statistik. H. Kungl. Majt:s befallningshafvandes femårsberättelser. Ny följd. Landshöfdinge-embetets uti Jämtlands län underdåniga berättelse för åren 1856–1860.* Stockholm.

Björklund, G. 1808. *Kokbok för husmödrar: innehållande beskrifningar öfver mer än 2000 anrättningar.* Stockholm.

Bonow, M. and Svanberg, I. 2011. »Säj får jag dig bjuda ur sumpen en sprittande ruda«: en bortglömde läckerhet från gångna tiders prästgårdskök, pp. 147–169 in M. Bonow and P. Rytkönen (eds.) *Gastronomins (politiska) geografi.* Stockholm.

Bonow, M. and Svanberg, I. 2012. Uppländska ruddammar: ett bidrag till akvakulturens kulturhistoria. *Uppland: årsbok* 2012: 123–152.

Bonow, M. and Svanberg, I. 2015. Urbana fiskdammar i 1600- och 1700-talets Sverige: strödda notiser om akvakultur i stadsmiljö. *Rig* 97: 215–222.

Bonow, M. and Svanberg, I. 2016. Monastiska fiskdammar i det senmedeltida Sverige, pp. 266–284 in M. Gröntoft (eds.) *Biskop Brasks måltider: svensk mat mellan medeltid och renässans.* Stockholm.

Brøndegaard, V. J. 1985. *Folk og fauna: dansk etnozoologi* vol. 1. København.

Bunting, S. B. and Little, D. C. 2005. The emergence of urban aquaculture in Europe, pp. 119–135 in B. Costa-Peire, A. Disbonnet, P. Edwards and A. Baker (eds.) *Urban Aquaculture.* Wallingford.

Carleson, C. 1769. *Hushåls-lexicon, hwaruti Det förnämsta, som angår Jordens behöriga skiötande, efter Alphabetisk Ordning, är sammanletat.* Nyköping.

Cawallin, S. 1857. *Lunds Stifts Herdaminne* vol. 4. Lund.

Cederlöf, O. 1766. *Oeconomiska Anmärkningar vid Skånska Karp-Dammar.* Lund.

Cederström, G. C. 1857. *Fiskodling ock Sveriges fiskerier.* Stockholm.

Cnattingius, B. 1932. Prästgårdar på Vikbolandet. *Lustgården* 13: 25–52.

Edling, A. 1910. *Ugglehultsboken: Carl M. Petersons småbruksdrift och föreläsningar.* Stockholm.

Ekre, R. 2007. Klostret i Lödöse, pp. 109–158 in J. Hagberg (ed.) *Kloster och klosterliv i det medeltida Skara stift.* Skara.

Ekström, C. U. 1831. Fiskarne i Mörkö Skärgård. *Kongl. Vetenskaps-Academiens Handlingar, för år 1830*: 142–204.

Ellenius, A. 1967. Konst och miljö, pp. 41–94 in S. Dahlgren (ed.) *Kultur och samhälle i stormaktstidens Sverige.* Stockholm.

Ericus, I. 1683. *M. Joh. Coleri Oeconomia, thet är, Hushåldz vnderwiijsning.* Stockholm.

Fagan, B. M. 2007. *Fish on Friday: Feasting, Fasting, and the Discovery of the New World.* New York.

Fischerström, J. 1761. Anmärkningar om Södra-Halland. *Kongl. Vetenskapsacademiens Handlingar* 22: 230–280.

Fischerström, J. 1785. *Utkast til Beskrifning om Mälaren.* Stockholm.

Flinck, M. 1996. *Tusen år i trädgården: från sörmländska herrgårdar och bakgårdar.* Stockholm.

Gadd, C.-J. 2011. The agricultural revolution in Sweden: 1700–1870, pp. 118–165 in J. Myrdal and M. Morell (eds.), *The Agrarian History of Sweden: from 4000 BC to AD 2000.* Lund.

Gardberg, C. J. 1959. *Åbo slott under den äldre vasatiden.* Helsingfors.

Gislén, T. and H. Kauri 1959. *Zoogeography of the Swedish Amphibians and Reptiles with Notes on Their Growth and Geography.* Stockholm.

Granlund, J. 1983. *Peter Månssons Bondakonst.* Uppsala.

Granlund, V. 1876. Johan III:s byggnads- och befästningsföretag 2. Bref ur Riks-Registraturet, pp. 27–254 in C. Silfverstolpe (ed.) *Historiskt bibliotek* vol. 2. Stockholm.

Hallenborg, C. 1913. Anmärkningar till Carl von Linnés Skånska resa. *Historisk tidskrift för Skåneland* 4: 293–373.

Hausen, R. 1915. Ur gamla familjepapper: Tidsbilder från förra århundradet. *Svenska litteratursällskapets Förhandlingar och uppsatser* 13: 177–288.

Hellquist, E. 1939. *Svensk etymologisk ordbok.* Lund.

Hoffmann, R. C. 1995. Environmental change and the culture of common carp in Medieval Europe. *Guelph Ichtyology Review* 3: 57–85.

Hoffmann, R. C. 1996. Economic development and aquatic ecosystems in Medieval Europe. *American Historical Review* 101: 631–669.

Hofmeister, E. 2004. Fiskeproduktion ved de ferske vande – fra karper til regnbuger, pp. 75–86 in E. Hofmester (ed.) *De ferske vandes kulturhistorie i Danmark.* Silkeborg.

Hofrén, M. 1937. *Herrgårdar och boställen: en översikt över byggnadskultur och heminredning å Kalmar läns herrgårdar 1650–1850.* Stockholm.

Hollberg, A. 1896. *Husmanskost: en hjälpreda för sparsamma husmödrar efter mångårig erfarenhet författad.* Stockholm.

Huldén, J. J. 1957. *Om Ebba Brahe.* Jakobstad.

Hultman, E. 1913. *Ekenäs stads dombok 1623–1675,* vol. 1. Helsingfors.

Hülphers, A. 1783. *Samling til korta Beskrifningar öfwer Swenska städer. 2dra flocken om städerne i Södermanland.* Westerås.

Johansson, H. 1964. *Ritus Cisterciensis. Studier i de svenska cisterciensklostrens liturgi.* Lund.

Jonsson, L. 1984. Djuren i staden. *Upplands fornminnesförenings tidskrift* 50: 88–94.

Juhlin-Danfelt, H. 1925. *Lantbrukets historia: världshistorisk översikt av lantbrukets och lantmannalivets utveckling.* Stockholm.

Kendla, M. 2000. Fischbenennungen auf Saaremaa und Muhumaa. *Linguistica Uralica* 3: 178–193.

Klemming, G. E. 1866. *Svenska medeltidskrönikor* vol. 2. *Nya eller Karlskrönikan.* Stockholm.

Knutberg, C. 1768. Sammandrag af de fyra till Kongl. Academien inkomna svaren på frågan om bästa sättet att inrätta och underhålla fiske-dammar, med egna tillägningar. *Kongl. Vetenskaps-Academiens handlingar* 29: 166–175.

König, C. H. 1757. *Inledning til Mecaniken och Bygnings-Konsten.* Stockholm.

Kornhall, D. 1968. *Sydsvenska fisknamn.* Lund.

Kungl. Maj. 14 November 1766 = *Kongl. Maj:ts Nådige Allmänne Stadga Och Ordning, För Rikets Hafs- Skär- Ström- och Insjö Fiske. Gifwen Stockholm i Råd-Cammaren then 14. Novembr. 1766.* Stockholm.

Larsson, J. 1922. Forellodling i dammar, pp. 660–706 in Nordquist, O. (ed.) *Sötvattensfiske och fiskodling. Svenska Jordbrukets Bok.* Stockholm.

Lenander-Fällström, A-M. 1987. Kvinnor i lokalhistoriskt perspektiv. Levnadsvillkor i Örebro vid 1600-talets mitt, pp. 108–119 in B. Sawyer and A. Göransson (eds.) *Manliga strukturer och kvinnliga strategier: en bok till Gunhild Kyle December 1987.* Göteborg.

Lepiksaar, J. 1965. Djurrester från det medeltida Ny Varberg. Fynd från karmeliterklostret ca 1470–1530. *Varbergs Museums Årsbok* 16: 73–102.

Lepiksaar, J. 1969. Nytt om djur från det medeltida Varberg. V*arbergs museum. Årshok* 20: 37–68.

Lilljeborg, W. 1891. *Sveriges och Norges fauna. Fiskarne* vol. 3. Upsala.

Lindgren, G. 1939. *Falbygden och dess närmaste omgivning vid 1600-talets mitt: en kulturgeografisk studie.* Uppsala.

Lindner, N. 1867. *Om allmogemålet i Södra Möre Härad af Kalmar län.* Uppsala.

Lindroth, S. 1967. *Kungl. Svenska Vetenskapsadademiens historia 1739 - 1818*, vol. 1. Stockholm.

Lindroth, S. 1975. *Svensk lärdomshistoria* vol. 3. *Frihetstiden.* Stockholm.

Lindstedt, P. 1912. *Om byggandet och skötseln af karp- och sutaredammar.* Lund.

Linnaeus, C. 1751. *Skånska Resa På Höga Öfwerhetens Befallning Förrättad År 1749.* Stockholm.

Linnaeus, C. 1755. *Flora svecica.* Stockholm.

Linnaeus, C. 1764. *Dissertatio medico-chirurgica de hirudine.* Upsala.

Linnaeus, C. 1899. *Hortus Uplandicus.* Uppsala.

Lönnberg, E. 1913. *Linnés föreläsningar öfver djurriket.* Uppsala.

Lönnqvist, B. 1993. Måltid och minne, pp. 9–33 in G. de la Chapelle and K. Hagelstam (eds.) *Bord - duka dig! Herrgårdsmat i Finland.* Helsingfors.

Lundberg, S. and Svanberg, I. 2010. Stone loach in Stockholm, Sweden and the royal fish in the seventeenth and eighteenth centuries Sweden. *Archives of natural history* 37: 150–160.

Lundberg, S. and Svanberg, I. 2016. European weather loach (*Misgurnus fossilis*) at Ulriksdal Palace, Stockholm, in the 1750s. *Archives of Natural History* 43: 163–166.

Makowiecki, D. 2008. Exploitation of early medieval aquatic environments in Poland and other Baltic Sea Countries: an archaeozoological consideration, pp. 753–777 in *L'Acqua Nei Secoli Altomedievali. Spoleto, 12–17 Aprile 2007.* Spoleto.

Malm, A. W. 1863. Svenska iglar, Disciferae, afbildade efter lefvande exemplar. *Kungliga Vetenskaps- och Vitterhets-samhället i Göteborgs Handlingar* 8: 153–262.

Malmgren, A. J. 1863. *Kritisk öfversigt af Finlands Fisk-fauna.* Helsingfors.

Melfors, E. 1991. *Ivar Axelssons Totts räkenskapsbok för Gotland 1485–1487.* Visby.

von Möller, P. 1871. *Halländska herrgårdar.* Halmstad.

Moyle, P. B. 1997. The importance of historical perspective: fish introductions. *Fisheries* 22 (10): 14.

Nilsson, A. 1939. Fiskodling i Skåne i äldre tid. *Skånes hembygdsförbunds årsbok* 1939: 86–99.

Nilsson, P. 2010. *Bortom åker och äng: förekomst och betydelse av kvarnar, fiske, humle- och fruktodlingar enligt de äldre geometriska kartorna (ca 1630–1650).* Uppsala.

Nilsson, S. 1855. *Skandinavisk fauna* vol. 4. *Fiskarna.* Lund.

Nisser, W. 1927. Per Brahe d.y:s Bogesund. *Upplands Fornminnesförenings Tidskrift* 43: 26–42.

114

Norbäck, O. G. 1884. *Handledning i fiskevård och fiskeafvel.* Stockholm.

Nordeide, S. W. and Hufthammer, A. K. 2009. Fishponds as garden features: the example from the Archbishop's Palace, Trondheim. pp. 277–282 in J.-P. Moreland and A. M. Mercuri (eds.) *Plants and Culture: Seeds of the Cultural Heritage of Europe.* Ravello.

Nordqvist, H. 1922. Karp- och sutarodlingar i dammar. pp. 587–659 in O. Nordqvist (ed.) *Sötvattensfiske och fiskodling.* Stockholm.

Nordström, T. and Dahlander, M. 1908. *Örebro slotts byggnadshistoria.* Örebro.

Olaus Magnus 1555. *Historia de gentibus septentrionalibus.* Roma.

Osbeck, P. 1996. *Djur och natur i södra Halland under 1700-talet.* Halmstad.

Piironen, J. 1994. Finlands fiskar och deras huvudsakliga utbredningsområde, pp. 63–81 in J.-O. Frier (ed.) *Truede ferskvandsfiskearter i Norden.* København.

Rasmuseen, H. 1959. Fiskedamme og fiskeopdræt, sp. 307–209 in *Kulturhistoriskt lexikon för nordisk medeltid* vol. 4. Malmö.

Reuterholm, G. 1909. En midsommarfärd från Stockholm till Åbo 1777. *Svenska Turistföreningens Årsskrift* 1909: 131–138.

Rietz, J. E. 1867. *Svenskt dialektlexikon: ordbok öfver svenska allmogespråket.* Lund.

Roberg, L. 1951. Resa till Väddö 1712. *Lychnos* 1950–51: 182–204.

Rothof, L. W. 1762. *Hushålls-Magasin. Första Delen, om Hushålls-Ämnen Til Deras nytta, bruk och skada, Beskrefne Uti Oeconomiska Föreläsningar.* Skara.

Rudbeck, O. 1947. *Olaus Rudbecks Atlantica* vol. 3. Uppsala.

Rålamb, Å. 1690. *Adelig öfningz fiortonde tom, medh behörige kopparstycken.* Stockholm.

Rålamb, Å. 1691. *Adelig öfnings trettonde tom.* Stockholm.

Sarmela, M. 1994. *Suomen kansankulttuurin kartasto* vol. 2, *Suomen perinneatlas.* Helsinki.

Schött, H. 1914. *Östergötlands läns hushållningssällskaps historia* vol. 2. Linköping.

Schultze, S. T. 1778. *Den swenske fiskaren, eller wälment underrättelseom det i Swerige nu för tiden brukeliga fiskeri.* Stockholm.

Schwerin, H. H. 1932. *Skånska herrgårdar efter Roskildefreden: En konsthistorisk undersökning av den skånska herrgårdsarkitekturens utveckling efter provinsens övergång till Sverige och fram till det nittonde seklets inbrott.* Lund.

Scroderus, E. 1640. *Johannis Amos Comenii vpläste gyllene tungomåls dör: eller alle språks och wettskapers örtegårdh thet är: en geenstijgh, til at lära thet latiniske, sampt hwart och itt språk.* Stockholm.

Silfverstolpe, C. 1895. *Vadstena klosters uppbörds- och utgiftsbok 1539–1570.* Stockholm.

Söderwall, K. F. 1953. *Ordbok Öfver svenska medeltids-språket. Supplement.* Lund.

Sorbonius, F. 1845/1693. Fornminnen i Näsums socken uti Skåne (Handskr. Af år 1693). *Runa: Antiquarisk Tidskrift* 1845: 100–106.

Spegel, H. 1998/1685. *Samlade skrifter av Haquin Spegel. Första delen. Guds werk och hwila,* vol. 1. *Text.* Stockholm.

Ståhlberg, S. and Svanberg, I. 2011. Catching basking ide, *Leuciscus idus* (L.), in the Baltic Sea: Fishing and local knowledge in the Finnish and Swedish archipelagos. *Journal of Northern Studies* 5: 87–104.

Sundman, G. 1893. *Finlands fiskar målade efter naturen*. Helsingfors.

Svanberg, I. 2000. *Havsråttor, kuttluckor och rabboxar: folklig kunskap om fiskar i Norden*. Stockholm.

Svanberg, I. 2007. »Deras mistande rör mig så hierteligen«: Linné och hans sällskaps-djur. *Svenska Linnésällskapets Årsskrift* 2007: 7–104.

Svanberg, I., Bonow, M. and Olsén, H. 2012. Fish ponds in Scania, and Linnaeus's attempt to promote aquaculture in Sweden. *Svenska Linnésällskapets Årsskrift* 2012: 83–98.

Svanberg, I.and Cios, S. 2014. Petrus Magni and the history of freshwater aquaculture in the later Middle Ages. *Archives of Natural History* 41: 124–130.

Tiburtius, T. 1768. Försök Gjorde vid Fiskeplantering i små Skogs-Sjöar. *Kongl. Vetens-kapsacademiens Handlingar* 29: 30–39.

Tidström, A. 1891. *Resa i Halland, Scania och Blekinge År 1756 med rön och anmärk-ningar uti Oeconomien, Naturalier, Antiqviteter, Seder, Lefnads-sätt*. Lund.

Tidström, A. 1978. *Resor i Västergötland 1756 och 1760*. Uppsala.

Triewald, M. 1746. Rön huru Iglar och Fyrfotor kunna fördrifwas utur Rude-Dam-mar. *Kongl. Swenska Wetenskaps Academiens Handlingar* 7: 216–217.

Trolle-Bonde, C. J. 1905. *Trolleholm förr och nu* vol. 1. Lund.

Trybom, F. 1885. *Om karpodling i allmänhet och särskildt om den vid Gustafsborg i Kristianstads län: föredrag hållet i Landbruks-Akademien den 19 jan. 1885*. Stockholm.

Trybom, F. 1895. Våra insjöar och deras skatter. En berättelse från en del insjöar inom Örebro län, besökta sommaren 1894. *Örebro Läns Kongl. Hushållnings-Sällskaps Qvartals-Skrift* 4: 33–75.

Ulväng, G. *Herrgårdarnas historia. Arbete, liv och bebyggelse på uppländska herr-gårdar*. Uppsala.

Wallin, C. 1962. *Tommarps Urkundsbok 1085–1600: klostret, hospitalet, staden, sock-nen* vol. 1. Stockholm.

Warg, K. 1755. *Hjelpreda i hushållningen för unga fruentimber*. Stockholm.

Whitaker, I. S., Rao, J. Izadi, D. and Butler, P. E. 2004. Ancient origin of, and trends in the use of medicinal leeches throughout history. *British Journal of Oral and Maxillofacial Surgery* 42: 133–137.

Wikström, J. E. 1840. *Stockholms flora, eller korrt beskrifning af de vid Stockholm i vildt tillstånd förekommande växter*. Stockholm.

Wiktorsson, P.-A. 1978. Sockenbor och utsocknes jordägare i bevarade medeltids-handlingar, pp. 144–162 in A. Garberg (ed.) *Sköllestabygden: tre Närkessocknar* vol. 1. Hallsberg.

Figure 4.1: Crucian carp pond at Alvastra Abbey (Photo: Ingvar Svanberg, 2010)

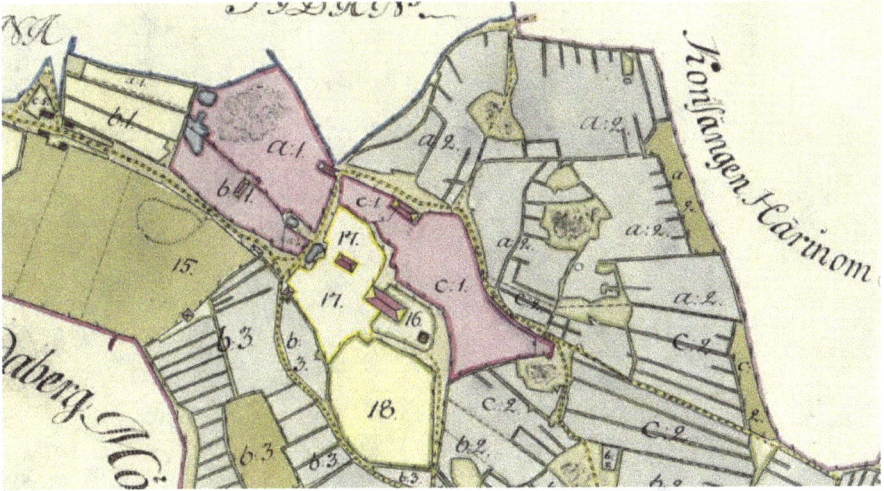

Figure 4.2: Cadastral map of Askeby from 1766

Figure 4.3: Pond variety of crucian carp by Wilhelm von Wright

Figure 4.4: Crucian carp pond at Ekolsund, Uppland.
(Photo: Ingvar Svanberg, 2012)

Figure 4.5: Former crucian carp pond at Sveaborg, Helsinki, Finland.
(Photo: Ingvar Svanberg, 2011)

CHAPTER 5

Fishponds and Aquaculture in Historical Times in Norway

Anne Karin Hufthammer and Dagfinn Moe

The origins and early history of aquaculture and fishponds in Norway are blurred. Only a few physical remnants (some pond constructions and a small number of common carp (*Cyprinus carpio* L.) or crucian carp (*Carassius carassius* (L.)), bones have been identified from archaeological excavations. Moreover, only some of the pond constructions and none of the bones have been dated. From early twentieth century Norway we know that both trout (*Salmo trutta* L.) and eel (*Anguilla anguilla* L.) have been kept in wells to keep the water clean. Historical records mention the keeping of bream (*Abramis brama* L.), eel and pike. In eastern Norway, by Lake Tyrifjorden, the practice was to lead bream from the lake into reservoirs where they were kept as a steady supply of fish for the summer. This tradetion dates back to time immemorial and ended in c. 1915 (Harsson 2000: 137).

Fishponds, whether they have been made for ornamental use or production reservoirs for the household, call for a certain level of knowledge with regard to both keeping fish and harvesting them. Furthermore, both the construction of the ponds and the keeping of fish are an investment that has to pay off; either as status symbols or through the production of necessary goods, i.e. to ensure the supply of fish.

The keeping of fish in ponds and reservoirs seems to have been limited to a few species in the past. In the translation into Danish of Max von dem Borne's book about fish breeding, ichthyologist Arthur Feddersen (1881) shows that the common fishpond species in Europe are in particular the common carp but also Northern pike (*Esox lucius*), zander (*Sander lucioperca*), European perch (*Perca fluviatilis*), tench (*Tinca tinca*), goldfish (*Carassius auratus*), ide (*Leuciscus idus*) and trout (*Salmo trutta*). However, the pond fish par excellence in Norway are cyprinids: the crucian carp, the common carp and the goldfish.

Carp and crucian carp in Norway

According to Steinar Kålås and Rune Johansen (1995), the carp was introduced to at least thirty-five locations between 1740 and 1992, mainly ponds and small lakes. These populations were probably extinct by the mid-1990s, but new translocations have taken place in recent years. Goldfish were previously found in several locations in Norway, but are currently found only in a small lake in south-eastern Norway (Huitfeldt-Kaas 1918; Mo 1996). The decline may be due to harsh climatic conditions or to destruction (filling in) of the ponds (Hesthagen and Sandlund 2007).

The oldest known Norwegian records that distinguish between the common carp and the crucian carps are Bishop Erik Pontoppidan's *Forsøk til Norges Naturlige Historie* written in 1752–3. Here he states that the carp is not a native Norwegian species and is therefore rare. Furthermore, he asserts that "*Karudse* (the crucian carp) are being kept, here as elsewhere, in lakes and ponds, both the large yellow (ones) and the smaller darkish (ones)". He also describes some extremely large specimens of crucian carp in a lake 3 miles further into the mountains from Lom, in the Gudbrandsdalen Valley (Pontoppidan 1753: 203).

In Norway there are few historical sources that mention fish-keeping in ponds and reservoirs. The oldest we know of is from the household of Olav Engelbrektsson, who was archbishop in Trondheim (Nidaros) from 1523 to 1537 Carps (*karusser* – here written *karudser*) as well as dry and fresh pikes are on the archbishop's menu for 1532 (Seip 1936: 1; Nordeide 2003b: 298). Generally, Norwegian historical records from the sixteenth and seventeenth centuries use only the word *karuss* and may be referring to the common carp, the crucian carp or both. It may be that in medieval times they did not differentiate between the two. It is therefore uncertain to which fish the plural form karuser in the archbishop's menu refers. If only one species is present, the most likely one would be the crucian carp.

It is highly likely that common carp were rare in Northern Europe in medieval times. For example, from northern Germany only two bones have been identified: one from the ninth or tenth century in Hitzaker and one from the twelfth century from Lübeck (Driesch 1982; Paul 1977). There are no medieval carp bones in Denmark, and only two from Leuuwarden in the Netherlands (Rosenlund 1976; Brinkhuizen 1983). It therefore seems that the carp was very rare in medieval times and was mainly introduced later in northern Europe. This assumption is supported by Otto Heuschmann (1957), who claims that the carp was introduced in Denmark as late as 1560. Further south, however, the carp was an important fish already in the fourteenth century. Historian Richard C. Hoffmann (1996), for instance, shows that records from the years 1338–75 document large quantities of carp being traded from Burgundy to the Rôhne and the Saône.

According to Eugene K. Balon (1995), the wild ancestors of the carp were distributed in the Black, Caspian and Aral Sea drainages and as far west as the Danube. In the postglacial period, crucian carp dispersed from eastern refugia to central and northern European waters (Libosvárský 1962). While the common carp is a southern species, the wild form of the crucian carp is endemic to northern Europe according to Wheeler (2000). Jiří Loborsvárský (1962) and Anton Lelek (1987: 343) claim a wider distribution that includes southern Europe. The modern distribution in Europe is discontinuous and in general restricted, but ranges from the Arctic Circle in Scandinavia to central France and the Black Sea in the south and from England in the west to the Lena River in the east (Loborsvarsky, 1962, Lelek, 1987: 343). The natural distribution of the fish, however, is uncertain (Holopainen et al. 1997). In the British Isles, it is probably only native to central and eastern England (Wheeler 2000).

In Norway, crucian carp are now found in many lakes connected to drainage systems in central-eastern Norway and in scattered lakes close to the southern and western coasts. The current theory is that the crucian carp dispersed naturally into eastern Norway (Figure 5.6) from freshwater drainage systems in southern Sweden at the time of the Ancylus Lake, i.e. 9–10,000 years ago (Øksnevad et al. 1995). A large number of prehistoric bone assemblages have been found in Norway. Many hold large quantities of fish bones, in particular from caves and rock-shelters in western Norway but also from open air sites in the north. None of these sites have produced bones of the crucian carp. This is a strong indication that the crucian carp never had its natural distribution in western and northern Norway. The same arguments are not valid for eastern Norway. From that region there are relatively few prehistoric assemblages and the preservation conditions for bones are poor, with extremely few bones of small species, i.e. fish. Thus, even though they never show up in the bone assemblages, the crucian carp may very well have been a part of the indigenous fish fauna of eastern Norway. The dispersal to coastal lakes in southern and western Norway is probably artificial and may date as early as the sixteenth century in some areas (Øksnevad et. al., 1995).

The distribution along the coast is somewhat patchy, with higher concentrations in the Bergen and Trondheim regions (Figure 5.6). This may be explained by the economic importance of these two towns in the past, Trondheim being the clerical and Bergen the trade centre of Norway in the Middle Ages. Both are thus the likely places for the introduction of new ideas and traditions, for example the keeping of fish in lakes and ponds.

Zooarchaeological evidence

In Norway, bone assemblages from approximately 140 archaeological excavations from the medieval period to the seventeenth century have been analysed. One assemblage may hold from a few to tens of thousands of identified bones. More than fifty of the materials are from urban locations, some 80 from rural areas and a few from clerical sites, i.e. monasteries and archbishops'/bishops' houses. From the medieval towns Oslo, Tønsberg, Stavanger, Bergen and Trondheim, a total of more than 326,000 mammal, bird and fish bones have been identified at the species level. The bones are mostly household remains and they provide a glimpse into the menus and activities of medieval life. All the studies demonstrate that fish, mainly marine species and in particular cod, played a central role in the medieval diet. In general, fish add up to 14–16 per cent of the identified bones from the medieval sites, but sometimes significantly more (Hufthammer 2003). This variation may be due to actual differences in the household economy, but may also be the result of differences in sampling techniques during excavation. Some of the bone assemblages have been collected by hand picking, others by the use of water sieving with meshes as small as 2 mm.

There are a number of studies on faunal remains from medieval Norwegian towns. From Oslo, Rolf Lie (1979, 1988, 1991) has studied bone assemblages from Oslogate 7, Mindets tomt and Søndre felt, Oslogate 4 and Kanslergaten. From Trondheim there are investigations from Folkebibliotekstomten (Lie 1989), Televerkstomten (Marthinussen 1992) and the Archbishop's Palace (Hufthammer 1999) and from Bergen from Rosenkrantzgate 4 (Wiig 1985) and Dreggen at the German Waft (Undheim 1987).

While saltwater fish are abundant, freshwater fish are rare in all assemblages, indicating that they played an insignificant, or rather another, role in the household than the marine fishes. For example, from the Archbishop's Palace in Trondheim, a total of 723 fish bones from the late sixteenth to mid-seventeenth centuries have been identified but only 6 are from freshwater fish: a cyprinid, a few Northern pike, and an eel (Hufthammer 1999). From the two medieval Trondheim town sites Televerkstomten (Marthinussen 1992) and Folkebibliotekstomten (Lie 1989), there are no freshwater fish, but more than 5,000 marine fish bones. The same paucity of freshwater fish is found in the other medieval town layers. Another example is in Oslo where there is one pike bone and 7,165 bones of marine fish from Mindets tomt (Lie 1988), and none from the Bishop's Palace (on file, University Museum of Bergen). Freshwater fish also seem to have been of little importance in post-medieval times. In seventeenth century sediments from Revierstredet they are absent (Lie 1981) and they are extremely rare in the assemblages from Kontraskjæret in the modern centre of Oslo (on file, University Museum of Bergen). All in all, based on the bone assemblages it is fair to claim that freshwater fish were not part of the regular diets of common citizens in medieval/early modern Norway.

However, the pattern is not absolute. In Oslo, from the thirteenth to six-teenth century layers from Kanslergaten tenth and eleventh to thirteenth century layers from Oslogate 4, eel is an important fish (Lie 1991). The Oslogate 4 site is next to an area that was inhabited by German shoemakers (Petter Molaug *pers comm.*) From Oslo there is another bone assemblage that obfus-cates the pattern of "no freshwater fish" even more: the remains from Nordre felt. In a small selection of bones from three trade houses next to the medieval town centre (the market square), large numbers of scales as well as a few bones of freshwater fish have been identified. There are scales of bream, one or several *Leuciscus* species, and perch. There are also cyprinid vertebrae, pharyngeal bones of the common rudd (*Scardinius erythropthalmus*) and of a *Leuciscus* species, and a great many bones of pike and perch. The presence of freshwater fish was nonetheless very much the exception rather than the rule.

Currie (1990) claims that in England fish were common to all men, but freshwater fish had higher status. Except for inland areas where freshwater fish were readily available, this may have been true in medieval Norway as well. The consumption of freshwater fish was probably for the few and either a signal of high status or based on religious regulations, i.e. fish consumption during fasting. However, the higher status of "freshwater fish consumption" must also depend on the exclusivity of the fish – and the taste and quality of the meat. Abundant and easily available fish would hardly gain such a status. Moreover, one would expect the status level to depend on how much effort was put into keeping and processing the fish.

The skeletal morphologies of the two carp species are very similar and in archaeological assemblages their bones are rarely identified at species level. However, they would have been identified as cyprinids, so identification problems are not the reason for their scarcity. The only sub-fossil remains identified with certainty as the common carp are from the Bygdøy kongsgård (Bygdøy royal estate) in eastern Norway, in total three bones: two pre-caudal vertebrae and a clavicula. The bones were found together in a clay layer and may well be from one fish, approximately 45 cm long. The impressive size indicates that the fish was quite old. The Bygdøy residence has been owned or used by the Norwegian Royal Family since 1305. Prior to that, the estate was owned by the "Maria" monastery that was established in 1147 at Hovedøya in the Oslo fjord. The oldest documentation of fishponds at the royal estate is from the seventeenth century, a time when there was a strong focus on fish breeding and the creation of elegant gardens (Berg 1952: 13–14).

However, it is from Trondheim, at the Archbishop's Palace, that we have the only find that verifies the presence of a fishpond in medieval Norway (Nordeide and Hufthammer 1993; Nordeide 2003b: 237–239). The arch-diocese was established in Trondheim (formerly Nidaros) in 1152 and remained the religious centre of Norway, Iceland, Greenland, the Isle of Man, the Hebrides, Orkneys, and Faeroes, as well as Jämtland, Härjedalen, and Bohuslän in Sweden until the Lutheran reformation in 1537, when it became one of many Danish-Norwegian royal estates. Until the archaeological excavations of the palace in the 1990s, little was known of the medieval buildings in the southern and eastern wings of the courtyard. During excavation it was discovered that the eastern wing of the precinct was a manu-facturing area holding a number of workshops associated with the arch-bishop's household and administrative functions, i.e. a bakery and a mint (Nordeide 2003a). At one stage, probably in the first third of the sixteenth century, there was an open area between the mint and a building of unknown function. In this area there was a rectangular construction approx. 13 m long and 4 m wide. In sediments from the bottom of the construction, eggs of water fleas, remains of water scavenger beetles, and spores of green algae were found, indicating that at some point the structure had been filled with fresh water. Moreover, a very well preserved rear half of a cyprinid fish was discovered in the basal silt; altogether 14 vertebrae, the anal fin, the back part of the dorsal fin and the tail fin – and a number of scales. Growth zones in scales and vertebrae indicate that the fish died in its fourth year, and the size of the vertebrae shows that it was 15–16 cm long (Figure 5.7).

Ponds and lakes

As fishponds are rarely mentioned in medieval Norwegian sources, information about the earliest history of this specialized method of food production and/or garden installations has to be found in regions that had close contact with Norway in medieval times. Great Britain is especially pertinent in this regard.

Currie (1984) claims that fishpond construction on a large scale did not begin until after AD 1066 in England, and that the earliest that are known after the Norman Conquest, at Foss outside York, were secular (Currie 1989). By 1175, a considerable network of fishponds had been established throughout the English shires and evidence of fishpond activity can be found amongst monastic records as well as through archaeological evidence. The earliest association of fishponds with monastic houses seems to be in the form of gifts from secular houses to monasteries (Currie 1989). According to Currie (1990), it seems that monastic fishponds were constructed in notable numbers from the early thirteenth century. On the other hand, Bond (1993) states that fishponds had become regular features of monastic precincts by the end of the twelfth century.

In Norway, a number of monasteries were established during the Middle Ages. The first was founded by the Benedictines c. AD 1100. Until the Lutheran reformation in 1536, when all monasteries were dissolved, a total of thirty monasteries were established, mostly along the coast from Østfold to Trøndelag (Lunde 1987). Many monasteries were established by the Benedictines and the Cistercians and both orders came from England (Gunnes 1996).

In England, ponds in medieval gardens almost always served as fishponds. Fish-keeping was generally organized with a breeding pond (the vivarium) and the smaller holding pond (the servatorium). The former were normally large, dammed features that were drained regularly so that the fish could be sorted. Fish that were selected for eating were transferred to the holding ponds (Currie 1990). The English ponds reared bream. According to Hoffmann (1994), carp, one of today's premier fishpond species, did not occur in twelfth and thirteenth century fishponds in England. Moreover, he hypothesizes that the manipulative, and eventually commercial, use of carp in medieval Europe was not emphasized until the mid-thirteenth century. Some of the earliest records of fish ponds in Norway are from the Trondheim region. There, in the first half of the sixteenth century, the influential Madam Ingerd of the estate Austrått, at the mouth of the Trondheim fjord, had a *crucian carp* pond (Ree and Wallem 1916; Nordeide 2000, p. 32, Nordeide and Hufthammer 2009). Madam Ingerd was contemporary to the Archbishop Olav Engelbrektsson who is known to have a fish pond by the castle the Archbishop arranged to be built at Steinvikholm (Nordeide and Hufthammer 1993). Furthermore Nordeide (*in prep.*) is interpreting a 2.5 m broad and 5.75 m long structure at the castle as a probable fishpond.

The structure with the half fish in the Archbishop's Palace is the only pond structure from medieval/early historical times that has been excavated in Norway. The central part of the pond was cut away by a more recent cellar but the rest of it was relatively well preserved (Nordeide 2003b: 237–239). The structure was a rectangular pit, 12.8 m long by a maximum of 4 m wide and 1.2 m deep. The walls were tightly lined with cut timbers in a traditional Norwegian building technique called *lafting* (log-cabin construction). The deeper parts were dug approx. 0.6 m down into the natural blue-grey clay. The timber frame rested on a shallow shelf running around the pond (Figure 5.4) and also enclosed a shallower section or platform at the northern end of the pond (Nordeide and Hufthammer 2009).

A number of engravings show that fishponds were also constructed in the trading town of Bergen shortly after the Lutheran Reformation. Based on maps and written sources, Harris (1991) has found at least twenty-three ponds in Bergen and its hinterland. The oldest records show four ponds from the late seventeenth century, two outside and two inside the protective walls of the royal estate of Bergenhus. These ponds seem to have been in use for a long time, at least one hundred years. One is even depicted on a map as late as 1848.

One of the ponds on the royal estate is unique in being, as far as is known, the first saltwater pond in Norway (Figure 5.2). On a map from 1679 it is depicted as a long, narrow pond outside the northern wall of Bergenhus, and on a map from 1712–7 the name *Bon. tel boe* is printed just above it (Bontlabo is the modern name of the seashore area close by). In c.1782 the pond is named the *Fisch-Teich* (Figure 5.2), and the drawings record the information that the water in the reservoir should be at the same level as the lower tide (Harris 1991: 34; Moe 2004).

In 1694 there was also a pond in the bishop's garden. Most of the ponds located in or near Bergen seem, however, to date from the mid-to-late eighteenth century. The ponds of three estates at some distance from the town may be somewhat older. The oldest records of fishponds at the Milde main estate south of Bergen are from 1719 (Figure 5.3). In medieval times, the estate belonged to the Dominican order. The records enumerate what is in the garden, and amongst wild trees, flowers and fruit trees, four fish parks (reservoirs/ponds) are mentioned. Moe (2004) suggests that they may have been constructed as early as 1679–81. Only one pond is mentioned in records from 1888, 1901 and 1915/1916 (Moe 2004). In 2006, a small excavation was conducted at the supposed location of the ponds and at a depth of 0.5–1.2 m a bottom with very solid clay-silt material and nicely made stone ridges made of pebbles were exposed (Moe et al. 2006). No fish remains were found and it is not possible to verify if it was a fishpond. However, both the location and construction (the hard packed material at the bottom of the structure) point in that direction. The pond would have been too small and shallow to be a breeding place for fish. The pond(s) must therefore have served as a servatorium for the estate kitchen (and probably also as an aesthetic element in the garden). Store Mildevatn (a lake that is located only a few hundred metres from the estate) is a likely candidate as the vivarium. Today the lake holds populations of crucian carp, ruffe (*Gymnocephalus cernuus*) and the common carp (Kålås 1995). All three species are quite rare in the region (it is the only population of ruffe in western Norway) and they were most certainly artificially introduced into the lake as their nearest natural occurrence is at least several hundred kilometres away (for the carp it would be thousands of kilometres).

Kålås and Johansen (1995) have made a review of the introduction of cyprinids and concomitant ponds in southern Norway. They list a total of 35 localities, of which 11 are named ponds. The majority of localities are from the Oslo region. Except for one introduction in Bergen in 1740, they all date from the nineteenth and twentieth centuries (Pontoppidan 1753: 203). However, new data from the Bygdøy royal estate near Oslo show a relatively early introduction of the Common carp in eastern Norway as well.

In eastern Norway, as in the Trondheim and Bergen regions, the medieval ponds seem with few exceptions to be connected to clerical and royal estates (Figure 5.1). From Oslo and nearby areas we know of fishponds from monasteries, i.e. from the Cistercians' Abbeys at Hovedøya (Figure 5.5) on the island of Hovedøya in the Oslofjord and the island of Tautra close to Trondheim there are remnants of fishponds (Fischer 1964; Ekroll 1996). At both abbeys there have been several ponds. Today, there are remnants of only one pond at Hovedøya. Originally it had a diameter of 18 metres and was 1.5 metres deep and was encircled by a row of stones, but is now only a depression in the ground (Foosnæs 2006). Though little archaeological data and no structural remnants are known, there are several claims of fishponds on other monasteries. Allegedly, in the early nineteenth century structural remnants from a possible fishpond were found at Munkeby monastery in Levanger, not far from Trondheim (Klüwer 1823 in Foosnæs 2006). A project conducted by graduate students in archaeology at the University of Bergen analysed the landscape around the Lyse monastery just south of Bergen and claimed to have identified a fishpond north of the abbey.[1]

Comparing the present distribution of the Crucian carp in Norway and the distribution of Catholic monasteries in medieval Norway, a striking pattern appears (Øksnevad et al. 1995; Hommedal 1999). The distribution of the fish overlaps with the distribution of the monasteries to a large degree. In our opinion, the pattern supports the long-held theory that Crucian carp were reared at Norwegian monasteries. Furthermore, as all monasteries were dissolved by the Lutheran reformation in 1556-7, this supports the notion that Crucian carp must have been one of the first (possibly *the* first) fish to be farmed in Norway. The fish was introduced into power centres in western Norway outside of its natural range. In eastern Norway, where the fish probably occurred naturally, the initiation of farming Crucian carp was merely an introduction of new ideas and traditions. Other fresh water fish were also kept in ponds, i.e. pike, bream and eel, but this is only reported from regions where they were naturally distributed. The tradition of keeping carp in ponds seems to be post-medieval and most likely originates in the eighteenth century in connection with the establishment of Renaissance gardens at noble estates and houses.

References

Balon, E. K. 1995. Origin and domestication of the wild carp, *Cyprinus carpio*: from Roman gourmets to the swimming flowers. *Aquaculture* 129: 3–48.

Berg, A. 1952. *Bygdøy Kongsgård. Haakon V – Haakon VII*. Oslo

Bond, C. J. 1993. Water managements in the urban monastery, pp. 43–74 in: R. Gildchrist and H. Mytum (eds.), *Advances in Monastic Archaeology*. Oxford.

Brinkhuizen, D. C. 1983. Visresten uit twee middeleeuwse vindplaatsen te Leeuwarden. pp. 19–20 in: W. van Zeist, R. Neef and D. C. Brinkhuizen, *Planten-, vis- en vogelresten uit vroeghistorisch Leeuwarden*. Leeuwarden.

Currie, C. K. 1984. Carp beginnings. *Carp Fisher* 7: 64–67.

Currie, C. K. 1989. The role of fishponds in the monastic economy, pp. 147–171 in: R. Gildchrist and H. Mytum (eds.), *The Archaeology of Rural Monasteries*. Oxford.

Currie, C. K. 1990. Fishponds as garden features, c. 1550–1750. *Garden History* 18: 22–46.

[1] http://www.hf.uib.no/arkeologisk/landskap/analyse.html.

von den Driech, A. 1982. Fischreste aus der slawisch-deutchen Fürstenburg auf dem Weinberg in Hitzacker (Elbe). *Neue Ausgrabungen und Forschungen in Niedersachen* 15: 395–423.

Ekroll, Ø. 1996. Tautra-ein del av Europas første industrikompleks. *Spor: Fortidsnytt fra Midt-Norge* 21: 36–38.

Espeland, E. 2004. Spor i jord. Projekt på Bogstad, pp. 11–14 in E. Espeland and L.-E. Mikkelsen (eds.), *Spor i jord. Parke og hagen på Bogstad restaureres.* Oslo.

Feddersen, A. 1881. *Fiskeavlen – Efter M.v.d. Borne: "Die Fischzucht".* København.

Fischer, G. 1964. *Klosteret på Hovedøya. Et cistercienseranlegg* (= Fortidsminner 61). Oslo.

Foosnæs, K. 2006. *Quod superest monasterii hic quondam fundati. En bygnigsarkeologisk undersøkelse av Munkeby Kloster.* Master thesis in Archaeology, Norwegian University of Science and Technology, Trondheim.

Gunnes, E. 1995. Ordner og klostre i norsk samfunnsliv. *Collegium medievale: tverrfaglig tidsskrift for middelalderforskning* 8: 131–145.

Harris, C.1991. *Bergen i kart fra 1646 til vårt århundre.* Bergen.

Harsson, M. 2000. *Stein – en storgård på Ringerike.* Hole.

Hesthagen, T. and Sandlund, O. T. 2007. Non-native freshwater fishes in Norway: history, consequences and perspectives. *Journal of Fish Biology* 71: 173–183.

Heuschmann, O. 1957. Die Weissfische (Cyprinidae), pp. 23–199 in: R. Demoll and H.N. Maier (eds.), *Handbuch der Binnenfischerei Mitteleuropas,* vol 3A. Stuttgart.

Hoffmann, R. C. 1994. Remains and verbal evidence of carp (*Cyprinus carpio*) in medieval Europe, pp. 139–150 in W. van Neer (ed.) *Fish Exploitation in the Past. Proceedings of the 7th meeting of the ICAZ fish remains working group.* (= Annales du Musee Royal de l'Afrique Centrale, Sciences Zoologiques no. 274). Tervuren.

Holopainen, I. J., Tonn, W.M. and Paszkowski, C. A. 1997. Tales of two fish: the dichotomous biology of crucian carp (*Carassius carassius* (L.)) in northern Europe *Annales Zoologici Fennici* 34: 1–22.

Hommedal, A.T. 1999. Kva fortel bygningsrestane av dei norske klostra om kontinental norm og norsk praksis innan ordenslivet?, pp. 149–183 in E. Mundal and I. Øye (eds.). *Norm og praksis i middelaldersamfunnet* (= Kulturtekster 14). Bergen.

Hommedal, A. T. 2011. Ora et labora – be og arbeid. *Årbok for Universitetsmuseet i Bergen* 2011: 35–49

Hufthammer, A. K. 2003. Med kjøtt og fisk på menyen. pp. 182–196 in O. Skevik (ed.), *Middelaldergården i Trøndelag. Foredrag fra to seminarer.* Verdal.

Hufthammer, A. K. 1999. Kosthold og erverv i Erkebispegården. En osteologisk analyse. Utgravningene i Erkebispegården i Trondheim. *NIKU Temahefte* 17. 47 p.

Huitfeldt-Kaas, H. 1918. *Ferskvandsfiskenes utbredelse og innvandring i Norge med et tillæg om krebsen.* Kristiania.

Kålås, S. 1995. The ecology of ruffe, *Gymnocephalus cernuus* (Pisces: Percidae) introduced to Mildevatn, western Norway. *Environmental Biology of Fishes* 42: 219–232.

Kålås, S. and Johansen, R. 1995. The common carp (*Cyprinus carpio* L.) in Norway. *Fauna norvegica Serie A* 16: 19–28.

Kleiven, E. and Hesthagen, T. 2012. *Fremmede fiskearter i ferskvann i Aust-Agder. Historikk, status og konsekvenser.* (= NINA Rapport 669). Oslo.

Lelek, A. 1987. *The Freshwater Fishes of Europe* vol. 9. *Threatened Fishes of Europe.* Wiesbaden.

Libosvárský, J. 1961. Zur palaeoborealen Verbreitung der Gattung *Carrassius* Jarocki, 1822. *Zoologische Jahrbücher, Abteilung für Systematik Öekologie und Geographie der Tiere* 90: 197–210.

Lie, R. W. 1979. Osleologisk material fra "Oslogt 7", pp. 108–124 in E. Scheia (ed.) *De arkeologiske utgravninger i Gamlebyen, Oslo.* vol. 3. *Feltene Oslogate 3 og 7. Bebyggelsesrester og funngrupper.* Øvre Ervik.

Lie, R. W. 1981 Dyrebein. pp. 257–271 in: E. Scheia (ed.) *Fra Christianias bygrunn. Arkeologiske utgravninger fra Revierstredet 5-7,* Oslo.

Lie, R. W. 1988. Animal bones, pp. 153–196 in E. Scheia (ed.) *De arkeologiske utgravninger i Gamlebyen, Oslo.* vol. 5. *Mindets tomt-Søndre felt. Animal bones, moss-, plant-, insect- and parasite remains.* Øvre Ervik.

Lie, R. W. 1989. *Dyr i byen – en osteologisk analyse.* (= Meddelser nr 18 fra projektet Fortiden i Trondheims bygrunn: Folkebibliotekstomten.). Trondheim.

Lie, R. W. 1991. Dyrebein fra Oslogt 4 og Kanslergt. 10, pp. 63–84 in E. Schia (ed.) *De arkeologiske utgravninger i Gamlebyen, Oslo* vol. 10. *Grøftegravninger.* Øvre Ervik.

Lunde, Ø. 1987. Klosteranleggene i Norge. *Foreningen til norske fortidsminnesmerker bevaring. Årbok* 141: 85–119.

Marthinussen, K. L. 1992. Et osteologisk material fra Televerkstomten. Unpublished thesis, Zoological Museum, University of Bergen.

Mo, T. A. 1996. Stamnestjernet – en ny lokalitet for gullfisk og bendelmarken *Dilepis unilateralis. Fauna* 49: 70–74.

Moe, D. 2004. De gamle hagedammene i Bergen, hva vet vi om dem? *Årbok for Bergen Museum* 2004: 53–61.

Moe, D., Hufthammer, A. K., Indrelid, S. and Salvesen, P. H. 2006. New approaches to garden history; taxonomical, dendrological, pollen analytical and archaeological studies in a 17th century Renaissance garden at the Milde estate, Norway, pp. 221–247 in J.-P. Morel, J. T. Juan and J. C. Matamala (eds.) *The archaeology of crop fields and gardens.* Ravello.

Nordeide, S. W. 2000. Steinvikholm slott – på overgangen fra middelalder til nyere tid. *NIKU – Temahefte* 23: 1–82. Trondheim.

Nordereide, S. W. 2003a. The Archbishop's Palace in Trondheim. *Norwegian Archaeological Review* 36: 135–140.

Nordereide, S. W. 2003b. *Erkebispegården i Trondheim. Beste tomta i by'n.* Trondheim.

Nordeide, S. W. (in prep). Fisk, faste og gamle gåter i middelalderborger. Manuscipt.

Nordereide, S. W. and Hufthammer, A. K. 1993. Fiskedam i Erkebispegården i Trondheim. *Spor* 1993:1: 44–45.

Nordereide, S. W. and Hufthammer, A. K. 2009. Fishponds as garden features: the example from the Archbishop's Palace, Trondheim, pp. 277–282 in J.-P. Morel and A. M. Mercuri (eds.) *Plants and Culture: Seeds of the Cultural Heritage of Europe.* Ravello.

Øksnevad, S. A., Poleo, A.B.S., Østbye, K., Heibo, K., Andersen, R A. and Vøllestad, R A. 1995. En ny teori om karussens innvandring og utbredelse i Norge. *Fauna* 48: 123–127.

Paul, A. 1977. Knockenfunde aus dem mittellaterlich-neuzeitlichen Lübeck (Grabung Hundestrasse 9–11). Manuscript.

Pontoppidan, E. 1753. *Det første Forsøk paa Norges Naturlige Historie*, vol. 2. Kjøbenhavn.

Ree, L. H. and Wallem, F. B. 1916. *Østraat*, Trondhjem.

Rosenlund, K. 1976. *Catalogue of Subfossil Danish Vertebrates. Fishes.* København.

Seip, J. A. 1936. *Olav Engelbriktssons regnskapsbøker 1532–1538*, Oslo.

Underheim, P. 1982. Osteologisk materiale fra Dreggen. En økologisk studie fra middeladerens Bergen. Unpublished thesis. Museum of Zoology, University of Bergen.

Wallem, F. B. 1917. *Steinsvikholm: erkebisp Olav Engelbrektssøns faste slot.* Trondhjem.

Wheeler, A. 2000. Status of the crucian carp, *Carassius carassius* (L.), in the UK. *Fisheries Management and Ecology* 7: 315–322.

Wiig, Ø. 1981. Faunal remains from medieval Bergen. *Fauna norvegica Ser A*, 2: 34–40.

Figure 5.1: Map from 1773 of the Bergenhus castle, Bergen. Four fishponds can be seen: the salt-water reservoir to the left and two adjacent, parallel ponds to the right, one for eel and one for Crucian carp. Close to the main castle wall is a pond labelled Fisch-dam. (The map is drawn by Hans von Gottberzk) (Courtesy of Slottsbiblioteket (The Royal Library, Oslo)

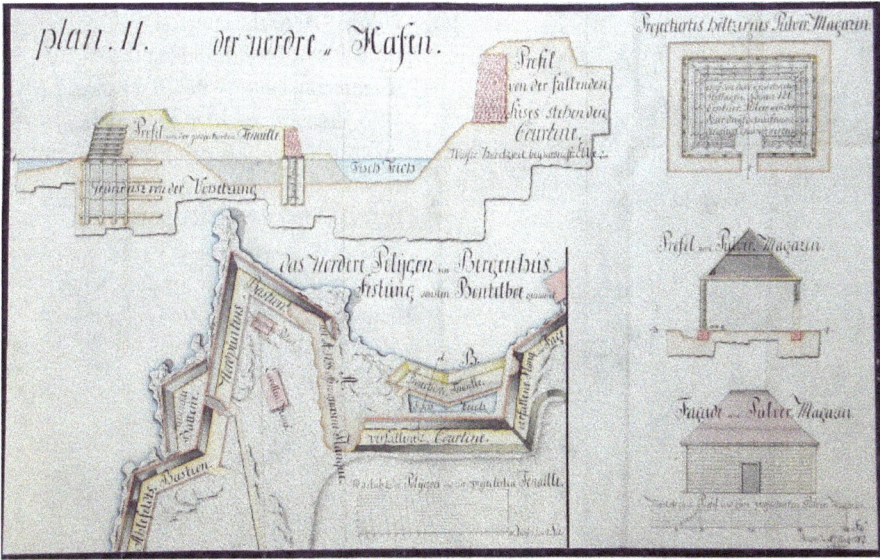

Figure 5.2: Illustration showing the salt fish reservoir. It is written that the water level in the pond should be kept at low tide level ("niedrigste Ebbe") (Moe 2004). The construction plan was drawn by Christoph Henrich Suckow in 1780 in connection with a military building project next to the reservoir). (Original at Riksarkivet)

Figure 5.3: Archaeological excavations at the Milde estate garden, in the area where records claim that there was a system of four fishponds. (Photo: Dagfinn Moe, 2006)

Figure 5.4: Archaeological excavation of a fishpond at the seventeenth century Bogstad Estate, Oslo. A square wooden construction, made in lafting (log-cabin construction) technique, was found in the central part of the pond. This construction was probably used for keeping the fish ready for the kitchen (Espeland 2004). (Photo: Dagfinn Moe, 2005)

Figure 5.5: The fishpond at Hovedøya monastery. (Photo: Alf Tore Hommedal, 2005)

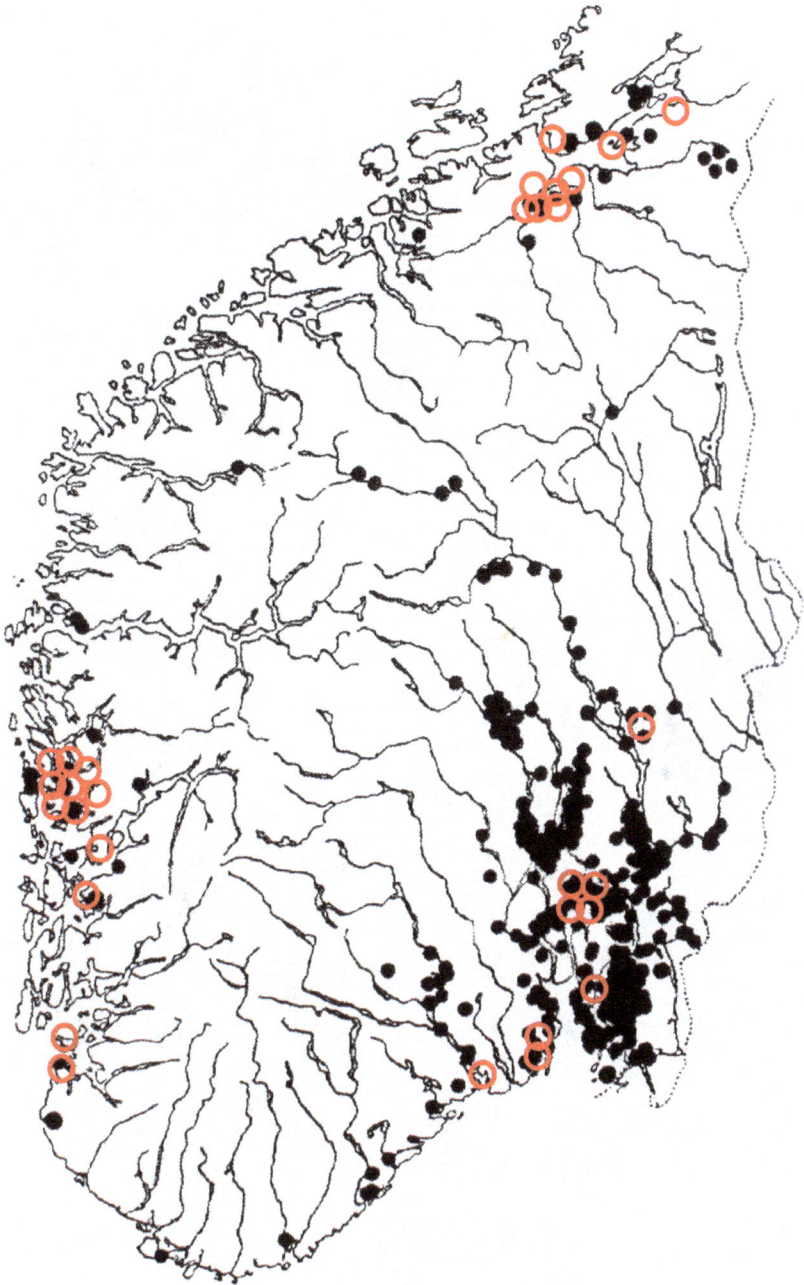

Figure 5.6: A map of the distribution of crucian carp (black dots) and medieval monasteries (red circles) in Norway.

Figure 5.7: The tail half of a cyprinid in situ from the fishpond at the Archbishop's Castle in Trondheim. (Photo: E. Baker, Directorate for Cultural Heritage)

CHAPTER 6

Fishponds in the Baltic States

Historical Cyprinid Culture in Estonia, Latvia and Lithuania

Madeleine Bonow, Stanisław Cios and Ingvar Svanberg

The three Baltic States – Estonia, Latvia and Lithuania – are currently among the smallest aquaculture producing countries in the European Union (Eurostat 2011: 142). The main species produced in Estonia is rainbow trout, *Oncorhynchus mykiss* (Walbaum, 1792), while the common carp, *Cyprinus carpio*, is the main species in Latvia and Lithuania. So far, there has been very little research into the history of fishpond culture in the region that today constitutes the Baltic States. However, the cultivation of cyprinids in ponds in this area can be traced to as far back to medieval times. As well as the common carp, the crucian carp, *Carassius carassius*, tench, *Tinca tinca*, and, more recently, during the Soviet-forced annexation, the Prussian carp, *Carassius gibelio*, have also been farmed. The latter species was introduced into the Baltic States in the late 1940s (Ojaveer, Pihu and Saat 2003: 231; Vetemaa et al. 2005; Aleksejevs and Birzaks 2011). Undertaking research across the Baltic region is complicated by the fact that the available sources are in many different languages, which is a consequence of the region's turbulent political history. Different rulers have conquered the area, and national borders have shifted numerous times. The present day borders of the three Baltic states of interest to this study emerged with the dissolution of the Soviet Union, when they reclaimed independence.

The historical establishment of pond culture in the Baltic territories is highly intertwined with the earlier monastic culture and feudal structures. For the moment though, we must be content with only a few examples until more in-depth studies become available. During 1627–8, when the northern parts of the Baltic territories were part of Sweden proper, Georg von Schwengeln (1590–1664), a Baltic-born German cartographer, was making maps of Estonia, Livonia and the island of Saaremaa (Ösel). He also drew the earliest scale maps of the greater Riga area and of Zemgale. In 1681, during the reign of the Swedish King Charles XI, thirty-eight surveyors, under the leadership of Arnold Emmerling, travelled to Riga to prepare cartographic material for the Great Land Cadastre. The mapping of Swedish Livonia took five years (Tarkiainen 2010). The map scale was in Swedish cubits, with the occasional addition of a diagonal scale. New cartographic methods and instruments ensured a greater degree of precision and quality than before, but they also removed marginal data (Sparītis 2009). Thanks to these historical maps we can describe and examine the pond culture prevailing in the seventeenth century. For the southern part of the Baltic territories (i.e. contemporary Lithuania), we rely very much on written documents (cf. Cios 2012). The material presented here should be regarded as a brief introduction to this rather neglected subject.

The monasteries

The oldest data about aquaculture is connected with the monastic orders, which played an important role in the Christianisation of the Baltic region, especially the Cistercians and Dominicans. The Cistercian Order held a crucial position up to the 1230s in integrating Livonia with the Christian world (Tamm 2009). They established the first monastery in Dünamünde (Daugavgrīva) in Livonia in 1205–7. According to Tuulse (1942: 268), the monastery was founded at the mouth of the Daugava River, where the proximity of the water made it possible for the monks not only to monitor this important waterway, but also to practice fish farming.

Very few reliable sources about monastic fishponds exist from Estonia, although some traces of fishponds are known. Both the Reval (contemporary Tallinn) and Riga Dominican monasteries were founded between the late 1220s and the early 1230s. These first monasteries have not survived and no evidence has been found of ponds in their vicinity. The St Bridgettine Convent in Pirita just outside Reval was a monastery for both monks and nuns. Its location was carefully chosen. It was situated in the harbour area, near the riverside lands, which was an area that functioned as an important trading site at the time. The Pirita Convent, which was based on St. Bridget monastery in Vadstena in Sweden, was founded in about 1400. The Teutonic Order played the main role in establishing the convent since it was on their land the buildings were constructed. In 1407, two monks from Vadstena Abbey arrived in Reval to counsel the merchants. The first permit to break dolomite to gather building material to build the complex was acquired in

1417. The abbey was consecrated in 1435 (Raam 1994; Tamm 2010; Markus 2013). On the old cadastral map from 1689 there is a pond in the close vicinity of the monastery, possible inspired by the ponds in Vadstena, but we have no record of what was kept in them.

Eight kilometres north of Tartu, there was a Cistercian monastery called Falckenau (Kärkna Abbey). It was founded in 1233 and destroyed in the Livonian war in 1558. According to Tuulse (1942), the location of the monastery is typically Cistercian, far from the main urban centre and the main roads, at the mouth of the river Amma. The high banks of the river provided the conditions for fish farming and the surrounding forest offered an opportunity for land clearance work. In the fourteenth century, the monastery still operated a mill, but the moat was by now largely overgrown and the fishponds were in disrepair (Tuulse 1942: 268). Despite this, on the 1783 cadastral map a pond is clearly visible near the monastery. However, it could have been made in the eighteenth century.

Another example of fishponds comes from Padise Cistercian Abbey in Hajdu County, which is said to have had a three-pond flow-through fish farming system. No local evidence exists and no medieval sources have been found to substantiate this. The first map of Padise is registered in the Swedish cadastre from 1697 and it indicates ponds connected with the manor house built in the vicinity of the convent (Ridbeck 2005). At the Kuimetsa Nunnery, fishponds have been found but they are not dated. On the map from 1687 there are two ponds on the manor's estate.

City ponds

A special feature in Swedish historical aquaculture was breeding crucian carp in ponds inside the cities. This is also known from the Baltic territories of Sweden proper. There were ten cities in Estonian territory in the late sixteenth century. After the wars of the early seventeenth century, town rights were preserved by Reval, Dorpat, Narva and (New) Pernau (Pärnu). Arensburg (Kuressaare) and Narva gradually regained their status; the rest had been destroyed or passed into private hands (Raun 2001). With 10,000 inhabitants Reval was the third largest Swedish city after Stockholm and Riga in the seventeenth century. The city had several ponds within the city walls. Written sources are few, but in *Revaler Kämmereibuch*, there is a passage dated 27 September 1460 which indicates that the Town Council paid half a Riga Mark for 300 crucian carp that were placed in St Gertrud's pond. The location of the pond is not clear, but a possible site of the pond is near the St

Gertrud Chapel by the Great Coastal Gate or Suur Rannavärav (Vogelsang 1976). All fish in the city's ponds were the property of Reval Town Hall. Fishing in the ponds was prohibited without prior consent in the form of a permit from the Council. A fine of one mark was the penalty for those who broke this law. There is also evidence of fish farming in the ponds on the 1688 and 1686 cadastral maps.

There are ponds located just outside the town centre on the 1729 Dorpat (now known as Tartu) map. According to zoologist, Benedict N. Dybowski crucian carp existed in all local rivers, on the banks of Lake Peipus, in several ponds on the Emajõgi River (Embach) and in the trenches of Dorpat. He remarks that strangely enough the species very rarely appeared in the fish market in Dorpat. Dybowski also claims that the crucian carp ponds were missing in Livonia's more remote areas (Dybowski, 1862:49). He states that there were small ponds almost everywhere in the town – in all the trenches and gardens, and in all other water reservoirs. He also observed that the fish grows very well, despite the water freezing during winter. There were not less than thirteen small ponds in the courtyard of Klattenberg's House on Aleksandri Street. There were also several ponds in the churchyard of the Old Believers Prayer House (Dybowski 1862: 50–52).

In Parnau (Pärnu) evidence has been found of ponds in gardens (*dyckgarten*). Town councillors Berndt Hessels, Luder Klanth, Melcher van Galen and merchant Johann (Hans) Sack all had gardens with ponds. It is assumed that these ponds were not merely dug for ornamental reasons, but for fish farming, although a fishpond is specifically mentioned in the town register only once. Melcher van Galen constructed ponds in connection with a sauna, which was common at the time (Põltsam and Vunk 2001). In some of the fortified towns there were also ponds, for instance in Lihula 1683.

Manorial pond culture during Swedish rule

Most Livonian lands were in private ownership by the end of the sixteenth century, and the owners were mostly Baltic Germans. The German and Danish vassals initially lived mostly in new towns and castles. Their country houses served as a stopover while collecting taxes and did not much differ from wealthier farms. The first vassals began moving from castles to manor houses in the late thirteenth and early fourteenth centuries (Beerencreutz 1997: 22–24).

The countryside had characteristic features influenced by German colonial culture. This consisted of towns and medieval urban structures with castles, manors, inns, mills, and dispersed peasant farmsteads (Sparitis 2009). During the Swedish regime 1561–1710 (formally at the Treaty of Nyland 1721), a manorial upswing with the formation of small estates was seen in the countryside (Beerencreutz 1997: 53). In the century of Swedish rule, rural manors were usually enclosed constructions. The main buildings of the manor were often placed around a courtyard. The rest of the buildings were distributed following the contours of the fencing around the territory, sometimes also at the sides of access roads (Sparitis 2009).

The rural population was now divided into two main groups: landed gentry and peasants. The landed gentry consisted of Germans and increaseingly of Swedes, while the peasants were Estonian-speakers and a tiny group of Swedish-speaking coastal dwellers (Raun 2001). At the end of the seventeenth century there were about 500 manors in Estonia and about one third of these were established in the first half of the century (Beerencreutz 1997: 55). In total, 1,254 manors have existed over the centuries. There are few written sources available that refer to the manorial pond culture. But in many of the Teutonic Order castles, including those in Reval, *Fischmeisters* were employed. A *Fischmeister* (Fish Master) is a lower officer of the Order. Manor earnings received from fishing and fish farming were administered by the *Fischmeister* (Turnbull 2003: 23). This suggests that these manors were involved in some form of cultivation or harvesting of fish, although it is more difficult to say from this equivocal evidence what the role of fishponds was in this enterprise.

Tuulse (1945) refers to fishponds in castles in Livonia as early as medieval times. On the east coast of Saarenmaa, Maasilinn Castle (Soneburg) was erected in the fourteenth century and subsequently reconstructed in 1518. The castle had a trench system and on the south side three large rectangular fishponds were created, and according to Tuulse (1942: 187) they were in the same form as at other waterfront castles around the country at that time. Sesswegen Castle (Cesvaine), located in the so-called Latvian Part of the Archdiocese, was built in the seventeenth century (Tuulse 1942: 205). The castle was not one of the Orders' residences, rather it belonged to Vilhelm Fredrik Taube, Fier and Sesswegen. The castle, however, was confiscated by King Gustavus Adolphus and was bestowed to Count Nils Brahe. The fortified castle had a central location in a densely populated area. On the basis of eighteenth century drawings and plans of Sesswegen one can see that apart from the protection offered by the wall and water surrounding the castle there are two large fishponds situated outside the castle walls (Tuulse 1942: 205). The Wesenberg (Rakvere) and Loodh castles had crucian carp ponds that are depicted on the 1683 map. On the sixteenth century map of Kremon Castle belonging to the Domecapitul, a large pond outside the castle can be seen that has a similar appearance to ponds at Swedish castles. In 1680, there are plenty of fishponds depicted on the map of Ronneburg Castle (Rauna), which was the residence of Riga's Archbishop.

King Gustav II Adolf's Field Marshal Gustav K. Horn was Governor General of Livonia and he wanted to build a residence for himself on his estate at Vainiži (Wainsel) near Limbaži (Lemsal). Horn, one of the king's closest companions-in-arms in the Thirty Years War (1618–1648), had participated in the conquest of Livonia and the king awarded him the districts of Alūksne (Marienburg) and Gulbene (Schwanensee), as well as the Vainiži estate, making Horn the third biggest land owner in Livonia after Axel von Oxenstierna and Banér. The surveyor Faber's layout plan (1649–54) of the manor ensemble and a Baroque garden are preserved in the National Archive in Stockholm. A new manor castle was planned that was surrounded by a circular moat and a fence. The plan was complemented with outbuildings, a mill, fruit and vegetable gardens and decorative parkland that included several large fishponds. This plan, however, was never realised (RA, 2025:02).

A map from 1690 shows a pond on the grounds of the Wrangelshof (a manor under Helmet). The castle had belonged to Axel Oxenstierna, who in 1636 gave it to Field Marshal Herman Wrangel in exchange for Wohlfahrt in the Livland region. Fishponds are shown on a 1689 map of Wiems (Viimsi) Manor. Additionally, Swedish style fishponds are shown on maps of Jegelecht (1692), Kuimetsa Castle (1687, 1765) and Saren Hoff (1688). Other seventeenth century estates with large ponds are Ruttigferhoff (1690), Rathshoff (1684) and Hofwet Viol (1703).

Fishponds at Estonian and Livonian manors after 1710

The 1710 capitulations of the corporations of knights and towns, confirmed by subsequent tsars until Aleksander II (1855–1881), established the relationships between Estonia, Livonia and the Russian Empire. Peter the Great gave the Baltic German nobility back their manor houses which the state had expropriated during the Swedish years. During all this time Estonia belonged to the German cultural sphere and this was further facilitated by the large-scale immigration of the German intelligentsia in the eighteenth century (Raun 2001: 37ff).

A large number of estates were beginning to rearrange their estates and gardens during this time and a number of ponds were constructed. Some of these newly constructed ponds were extremely elaborate, for example Pöllküll (1882), Kattentack, Neu-Oberpahlen and Wannamois. Anzen (Antsla) Manor was one of the largest fish producers in Estonia during this period and it also operated as a provider of breeding stock to all the others. The general understanding is that it is likely that aquaculture in one form or another was practised here from the Middle Ages when the owners were the noble family von Uexkülls.

Another manor that practised fish farming during this time was the Piirsalu Estate, during the life-time of Cornelius von zur Mühlen (1756–1815), whose favourite hobby was fish farming. All of the following estates had large ponds depicted on maps: Addaffer, Annigfer, Chatarinenhof Hohensee, Harju-Madise Church Manor, Hermannshof, Kadrina, Kattentack, Kawast, the Karkus Castle, Laimetz, Poll and Neu-Poll, Pöllküll, Menan, Metzikus, Neu-Oberpahlen, Saku, Saue, Sutlep, Vääna, Wallkull, Wannamois and Seyer and Äntu Estate.

Carp ponds on Lithuanian estates

Carp production seems to have been known in the southern Baltic region since late medieval times. In 1402, the inhabitants of Klaipeda Castle consumed 28,000 carp. This tells us that Lithuanian fishpond culture dates back at least to the fourteenth century. Such significant fish consumption indicates that pond culture was well-developed in the Klaipeda region during this time (Žulkus and Daugnora 2009).

The first references to fishponds are in the Statute of the Grand Duchy of Lithuania from 1529 (Czacki 1861, II: 201, 254, 286). The statute refers to how the construction of ponds (and mills) on streams should not inundate other mills or meadows and to punishments for thieves fishing in ponds. Under the statute, when a thief is caught for the first time he is whipped, for the second time his ear will be cut off, and for the third time he will be treated like a thief, i.e. executed. Such references to ponds, as well as the types of punishment, indicate that in the sixteenth century both feudal lords and monarchs were strongly economically engaged in fishpond culture.

Of particular interest are references to carp related to the aristocratic Radziwiłł family. Although the main residence of the family was in Nieświez (Nesvizh in Belarus), the aristocrats owned considerable estates across the whole territory of the Grand Duchy.

The first reference to the Radziwiłł family is in a letter written in 1567 by King Sigismund Augustus to Mikołaj "Sierotka" Radziwiłł, in which the king expressed his hope that during his stay in the Lebedzev estate the aristocrat would provide fish from ponds for the royal table (Kaniewska 1999: 537). The next reference is by Jankowski (1898), who states in his description of the Oszmiana (Ašmena) district in southern Lithuania that in the seventeenth century in Lubcza (Lubča by the Niemen River/Nemunas), the Radziwiłł family had "five ponds which provided several fish, in particular carp" (Jankowski 1898: 104, 147). He also presents a letter written on 21 November 1758 in Worończa (Varonča) by Józef Niesiołowski, in which he states, "I will give an order to catch carp and send the fish on a good road, so they will not get injured". Carp are also mentioned in Hieronim Florian Radziwiłł's 1747 diary (Radziwiłł 1998: 29, 35, 45) with reference to his estates in Wyzna and Niehniewicze. The diary describes how Hieronim Florian Radziwiłł as a form of recreation spent some time by the ponds, observing fishing with a seine net, and how he personally took larger carp from the net, while releasing the smaller ones. It also recounts his strong interest in the management of ponds when he gives orders to clear the ponds of weeds. A still greater interest in fishponds was shown by Udalryk Radziwiłł, who invented a machine to cut reeds and other aquatic vegetation in order to increase the productivity of the ponds (Radziwiłł 1761; Górzyński 1964). However, he was criticized by others, who thought he should be doing other things with his time (Bagiński 1854: 36–7). Finally, in a letter from 1902, Michał Abłamowicz mentions "large ponds well stocked with carp", owned by the Radziwiłłs in Nieśwież (*Fałat...* 2008, II: 321).

There are also two sources that indicate consumption of carp by the Radziwiłłs. The first source is a family cookbook dating from 1686–1688 (Moda... 2011). Carp is mentioned in five recipes (for comparison: pike in eighteen, eel in two, herring and perch in one). This indicates that carp was an important culinary fish on the aristocratic menu. The second source is a story by Rzewuski (2000: 174), first published in 1845, in which carp with sweet honey sauce is mentioned as a dish on the menu of a feast. Historically this dish has been the most popular carp dish in Poland.

All of the accounts described above indicate that the Radziwiłł family owned carp ponds for almost 400 years. It is also likely that most of the carp cultivated ended up on the aristocrats' and guests' own table.

There are more sources that indicate a pond culture during this time. In a description of Szkudy (Skuodas) in north-west Lithuania, Potocki (1874: 245) states that in 1793 a rich owner of an inn boasted that he had various fish stocks in ponds for his guests – carp, tench, crucian carp, pike and perch. Morawski (1858: 82), describing the village Ustronie (Jundeliškės) by the Wierzchnia River (Verkne), a tributary of the River Niemen in southern

Lithuania, states that "in the old days there was a famous pond, which for over a century has been transformed into a meadow, in which previously carp were kept, most famous in Lithuania. Though at that time carp was not a rare fish in the region, these fish had a particular taste and reached such a size that – like whitefish in Lake Wigry [north-east Poland] – they were sent to the kings as a great delicacy". In his description of the life of nobility in the countryside in the region of Grodno (Belarus), Count Tyszkiewicz (1865: 7) observed that in one pond there were only carp, while in another only Crucian carp.

The information discussed here should by no means be considered exhaustive but it indicates that carp culture in Lithuania has a long tradition dating back to at least late medieval times. Carp cultivation appears to have been a cultural import related to the Teutonic Order, since the oldest sources of information on carp in Poland is from the northern part of the country (Joachim 1896). The Christianization of Lithuania in 1387 might also have served as an additional strong stimulus to consume more fish and relatedly to develop pond culture.

In Lithuania, in contrast to Poland, no large pond systems were developed, but rather the pond infrastructure consisted of small numbers of small ponds, which were owned mainly by the nobility. Furthermore, the fish cultivated were usually not destined for the market (this seems to be a modern development), but for the landlord's table.

Pond culture was well developed in the seventeenth and eighteenth centuries and this continued until 1861, when the Tsar's emancipation reform in Lithuania and Poland ended the serf system. In the new economic order, rising labour costs led quickly to the ruin of pond culture and by the end of the nineteenth century most ponds in Lithuania were in a very poor state (Staniewicz 1902).

Modernisation of aquaculture

The upper class and the state initiated liberation of peasants from serfdom in 1816 in Estonia and in 1819 in Livonia. The manor owners were compensated with land in return for abandoning their right to own the peasants. In the middle of the century new agrarian laws were passed. This laid the foundation for the purchase of farms and the emergence of peasant landowners and peasants began buying farmsteads from the estates at free market prices (Raun 2001: 45ff).

In November–December 1905, the Czarist government declared a state of war in the Baltic provinces. Within one week (12–20 December), the bands of workers and peasants, mostly in northern Estonia, destroyed, burnt down or looted about 160 Baltic German manor houses (i.e. every fifth manor). After the Russian revolution in February 1917, Estonians were integrated into one autonomous Estonian national province (Raun 2001: 82ff).

During this time, there were estates with large ponds and fish farming was carried on. One of these was the Löwenruh (Roosna) summer estate, which had been restored in the mid-nineteenth century. During the restoration the grounds were landscaped and nineteen fishponds were constructed. The fish were sold to the markets in Saint Petersburg. The manor was destroyed in a fire in the 1880s but fish production continued although not as extensively as earlier. Anna Graf sold the estate (26 ha.) in 1936 and fish farming on the estate was abandoned (Tõnuriste et al. 1976: 248).

According to Paaver et al. (2001), modern fish farming began in Estonia in the 1890s. They argue that it was German landowners who initiated farming of brown trout (*Salmo trutta fario*) and common carp (*Cyprinus carpio*) in ponds and developed it into a profitable branch of economy. Several trials were carried out by fish farm owners to improve the technology of pond farming. These trials included Staël von Holstein at the Antsla fish farm and Friedrich von Berg who owned the large Sangaste estate where he was experimenting with developing rye, potatoes, fish and horses, among other things. Berg wrote extensively about the problems of fish farming in ponds and his experience of solving them (Tohvert 1995).

Fish farming in Estonia more of less came to a halt from the period covering World War I to the end of World War II. (Paaver et al. 2001). That said, there are newspaper reports of some attempts to start fish farming, among them a salmon farming initiative in the pond "from the times of the barons" at Luke estate (*Postimees* 12.06.1935 nr. 157, p. 6). There was also a report that Professor Heinrich Koppel wanted to start fish farming in a lake at the Vissi Estate (*Postimees* 18.05.1935, nr. 134, p. 8.). Another attempt reported was an effort to start private fish farming of trout in the Elva Lakes (*Postimees*, 04.08.1934, nr. 210, p. 7, *Postimees*, 02.04.1935, nr.91, p. 6). All these endeavours ended with the onset of World War II and the occupation of Estonia by the Soviets.

Common carp is commonly farmed in contemporary Estonia. The largest fish farms are situated in the basin of the Emajõgi River, the Väike Emajõgi River and the Narva River. In 1990, which is considered to be the most productive year in Estonian aquaculture, 917 tons of carp were cultivated on fish farms. However, after liberation from the Soviet Union many fish farms ceased cultivating, although it has become popular with many farmers to keep carp in small ponds (Ojaveer, Pihu and Saat 2003: 239–240). Some attempts have been made recently to use aquaculture to cultivate other species such as grass carp, *Ctenopharyngodon idella* and bighead carp, *Aristichthys nobilis*. The most important taxon of aquaculture in Estonia is the rainbow trout, although production of this species has decreased in post-Soviet Estonia (Ojaveer, Pihu and Saat 2003: 114, 175, 240).

In Lithuania, a new chapter in the development of pond culture opened late in the nineteenth century. There were two leading pioneers at that time. One important actor was Mykolas Girdvainis (Michał Girdwoyń) (1841–1925), a fishery and aquaculture specialist who had visited several distinguished European scientific institutions (Gečys and Lirski 2011). He became famous not only in Lithuania and Poland but also throughout Europe, as he was responsible for establishing over 400 pond farms. Most of the farms established in Lithuania at that time were under his supervision. The aggregated water area for pond infrastructure amounted to more than 10,000 ha.

Girdvainis began to work in the fishery sector in the 1870s. It is likely that he constructed a fishpond in Verkiai, near Vilnius, at that time. This pond was regarded as a pioneer of modern design in the country. He designed carp and trout ponds on the Tyszkiewicz family estate in Waka (Vokė), near Vilnius, in around 1880–1885. These ponds, which still exist, were the first trout ponds established in Lithuania. Trout were sent to market, mainly to Saint Petersburg and Warsaw, less so to Vilnius. After marrying in 1885, Girdvainis moved to his wife's estate in Iszliny in western Lithuania, where he set up his private, well-known fish farm. He continued his work in this sector until 1916, when almost all the fishery institutions and enterprises he had established were destroyed during World War I. He was also the author of an 1881 pond culture manual written in Polish.

The second key actor was Cezarijus Stanevičius (Cezary Staniewicz) (1839–1909) who held a PhD and was a physician and ichthyologist. He was the first chairman of the Vilnius Section of the Imperial Fisheries Association in St Petersburg, established on 21 February 1901 (Anonymous 1904) and continued in this position until his death. His interest in fish farming stemmed from the fact that as a physician he noted the lack of proteins in the diet of his countrymen. Early in the twentieth century, Stanevičius published several works on fish and fisheries in Lithuania, including a 1902 work on pond culture (Staniewicz 1903). These books are extremely rare today.

References

Aleksejevs, E. and Birzaks, J. 2011. Long-term changes in the ichtyofauna of Latvia's inland waters. *Scientific Journal of Riga Technical University* 7: 9–18.

Adamson, A. 2009. Prelude to the birth of the "Kingdom of Livonia". *Acta Historica Tallinnensia* 14: 31–61.

Anonymous 1904. Stan rybactwa na Litwie. *Okólnik Rybacki* 72: 273–277.

Arbusow, L. 1919. *Wolter von Plettenberg und der Untergang des Deutschen Ordens in Preussen. Eine Studie aus der Reformationszeit Livland.* Leipzig.

[Bagiński, W. W. K.] 1854. *Rękopism X. Bagińskiego (1747–1784).* Wilno.

Berencreutz, M. 1997. *Gods och lantbönder i västra Estland.* Stockholm.

Bunting, S. B. and Little, D. C. 2005. The emergence of urban aquaculture in Europe, pp. 119–135 in B. Costa-Peire, A. Disbonnet, P. Edwards and A. Baker (eds.) *Urban Aquaculture.* Wallingford.

Cios, S. 2012. Chów karasia pospolitego *Carassius carassius* w Polsce od XVI do XIX wieku. *Forum Faunistyczne* 2 (1–2): 1–4.

Czacki, T. 1861. O litewskich i polskich prawach. Vols. I–II. Kraków.

Dybowski, B. N. 1862. *Versuch einer Monographie der Cyprinoiden Livlands: nebst einer synoptischen Aufzählung der Euopäischen Arten dieser Familie.* Dorpat.

Eurostat Pocketbook 2011. *Agriculture and fishery statistics. Main results – 2009–10.* Luxembourg.

Federlay, B. 1946. *Kungligt majestät, svenska kronan och Furstendömet Estland 1592–1600.* Helsingfors.

Gečys, V. and Lirski, A. 2011. Sto siedemdziesiąta rocznica urodzin słynnego ichtiologa i projektanta europejskich gospodarstw karpiowych i pstrągowych, inicjatora hodowli ryb na Litwie, Michała Girdwoynia. *Komunikaty Rybackie* 4: 26–29.

Górzyński, S. 1964. *Zarys historii rybołówstwa w dawnej Polsce.* Warszawa.

Joachim, E. (ed.) 1896. *Das Marienburger Tresslerbuch der Jahre 1399–1409.* Königsberg.

Kaniewska, I. (ed.). 1999. *Listy króla Zygmunta Augusta do Radziwiłłów.* Kraków.

Markus, K. 2013. The Pirita Convent in Tallinn: A Powerful Visual Symbol for the Self-Consciousness of the Birgittine Order. *Kungl. Vitterhets Historie och Antikvitets Akademien. Konferenser* 82: 95–110.

Mänd, A. and Randla, A. 2012. Sacred Space and Corporate Identity: The Black Heads' Chapels in Tallinn and Riga. *Baltic Journal of Art History* Autumn 2012: 43–80.

Ojaveer, E., Pihu, E. and Saat, T. 2003. *Fishes of Estonia.* Tallinn.

Oprawko, H. and Schuster, K. (eds.) 1971. *Lustracja województwa sandomierskiego 1660–1664.* Vol. I. Kraków.

Paaver, T., Tohvert, T. and Kangru, M. 2001. History of aquaculture research in Estonia. *Proceedings of the Estonian Academy of Sciences. Biology Ecology* 50: 211–221.

Pełczyński, M. 1960. Studia macaronica. Stanisław Orzelski na tle poezji makaronicznej w Polsce. *Prace Komisji Filologicznej, Poznańskie Towarzystwo Przyjaciół Nauk, Wydział Filologiczno-Filozoficzny* 20(1): 1–246.

Põltsam, I. and Vunk, A. (Hrsg.) 2001. *Quellen zur Geschichte der Stadt Pernau 13.-16. Jahrhunder* Bd. I. Pärnu.

Potocki, L. 1874. *Kazimierz z Truskowa czyli pierwszy i ostatni litewski powstaniec.* Poznań.

Potocki, W. 1915–1918. *Moralia.* Vols. 1–3. Kraków.

Prochaska, A. (ed.) 1892. *Archiwum domu Sapiehów.* Vol. 1. Lwów.

[Przetocki, H.] 1911. *Postny obiad abo zabaweczka wymyślona przez P.H.P.W.* Kraków.

Raam, V. 1993. Dominiiklaste Katariina klooster Vene t. 12–20. *Eesti Arhitektuur 1: Tallinn,* peatoim. V. Raam. Tallinn.

Raam, V. 1984. Das Birgitten-Kloster in Tallinn/Reval. Empore und Altäre. *Nordost-Archiv. Zeitschrift für Kulturgeschichte und Landeskunde* 75: 63–84.

Raam, V. and Tamm, J. 2006 *Pirita Convent: the History of the Construction and Research.* Tallinn.

Radziwiłł, H. F. 1998. *Diariusze i pisma różne.* Warszawa.

Radziwiłł, U. 1761. O wynalezieniu moim machyny, która na stawie płynąc, staw z trzciny, rogoziny, y wszelkiego wyczyszcza zielska, która machiną sarna dzika konsekrowała. *Nowe wiadomości ekonomiczne i uczone* 11: 685–696.

Raun, T. U. 2001. *Estonia and Estonians* Stanford, CA.

Rej, M. 1914. *Zwierciadło,* vols. 1–2. Kraków.

Ridbeck, H. 2005. *Padise läbi aegade.* Tallinn.

Rybarski, R. 1931. *Gospodarstwo Księstwa Oświęcimskiego w XVI wieku.* Kraków.

Rzewuski, H. 2000. *Listopad.* Kraków.

Schneider, G. 1925. Die Süsswasserfische des Ostbaltikums und ihre Verbreitung innerhalb des Gebietes. *Archiv für Hydrobiologie* 16: 133–155.

Sidrys, R. V. 1999. Fish names in the Eastern Baltic: etymology, ecology, economy. *Istorija* 41: 3–23.

Soom, A. 1949. Herrgården i Viimsi under senare hälften av 1600-talet. *Svio-Estonica* 9: 91–120.

Spārītis, O. 2009. Some Aspects of Cultural Interaction between Sweden and Latvian Part of Livonia in the 17th Century. *Baltic Journal of Art History Autumn 2009.*

Staniewicz, C. 1902. *Międzynarodowy kongres rybacki w Petersburgu.* Kraków.

Staniewicz, C. 1903. *Dział rybacki na wystawie rolniczej w Wilnie r. 1902.* Kraków.

Tamm, J. 2010. Residences of Abbesses in Estonian Monastic Architecture. pp. 31–35 in J. Tamm, *Eesti keskaegsed kloostrid*. Tallinn.

Tamm, M. 2009. Communicating Crusade: Livonian Mission and the Cistercian Network in the Thirteenth Century. *Ajalooline Ajakiri* 129/130: 341–372.

Tarkiainen, Ü. 2010. Die Vermessung Livlands. *Forschungen zur Baltischen Geschichte* 5: 59–74.

Tomczak, A. (ed.) 1963. *Lustracja województw wielkopolskich i kujawskich 1564–1565*. Vol. 2. Bydgoszcz.

Tõnurist, E., Järv, V. and Kahk, J. (eds.) 1976. *Sotsialistliku põllumajanduse areng Nõukogude Eestis: artiklite kogumik*. Tallinn.

Turnbull, S. 2003. *Tannenberg 1410: Disaster for the Teutonic Knights*. Oxford.

Tuulse, A. 1942. *Die Burgen in Estland und Lettland*. Dorpat.

Tyszkiewicz, E. 1865. *Obrazy domowego pożycia na Litwie*. Warszawa.

Willoweit, G. 1969. *Die Wirtschaftsgeschichte des Memellandes*. Vol. I. Wissenschaftliche Beiträge zur Geschichte und Landeskunde Ost-Mitteleuropas. Im Auftrage des Johann Gottfried Herder-Instituts (Hrg. E. Bahr). Nr. 85/1. Marburg/Lahn.

Vetemaa, M., Eschbaum, R., Albert, A. and Saat, T. 2005. Distribution, sex ratio and growth of *Carassius gibelio* (Bloch) in coastal and inland waters of Estonia (northeastern Baltic Sea). *Journal of Applied Ichtyology* 21: 287–291.

Vogelsang, R. 1976. *Kämmereibuch der Stadt Reval, 1432–1463*. Köln.

Wyczański, A. 1959. *Lustracje województwa lubelskiego 1565*. Wrocław-Warszawa.

Žulkus, V. and Daugnora, L. 2009. What did the Order's Brothers Eat in the Klaipėda Castle? The Historical and Zooarchaeological Data. *Archaeologia Baltica* 12: 74–87.

Archives

National Archives of Sweden, Stockholm

RA, 2025: 02, K. G. Lagerfelds gåva 1960, No. 2, Bl. 1–8, Livland. Wainsells slott, trädgård och ägor.

Military Archives of Sweden, Stockholm

Estonian digital active for maps: Copey der Geometrischen Charte von Stadt Dörpt Reference code EAA.995.1.6851 sheet 5; Wiems GenerL: GouverTZ: i Estland Tafelgodz1689 Reference code EAA.1.2.C-II-35

Geometrische Charte von dem publ. Guthe Falckenau und dem davon privat gewordenen Theile Marrama 1783 Reference code EAA.2072.3.39c sheet 1; Charte öffwer den Stridigheet Emellan Payushoff Bönder och Ruttigferhoff 1690.EAA.308.2.209; Geometrische Charte von dem im Rigaschen Gouvernement, Dorptschen Kreise und Koddaferschen Kirchspiele belegenen privaten

Gute Chatrinenhoff gemessen und eingetheilt im Jahre 1811 und 1812 EAA.1809.2.274 sheet 1.

Figure 6.1: The Pirita monastery 1689 (EAA1.2.C-II-35)

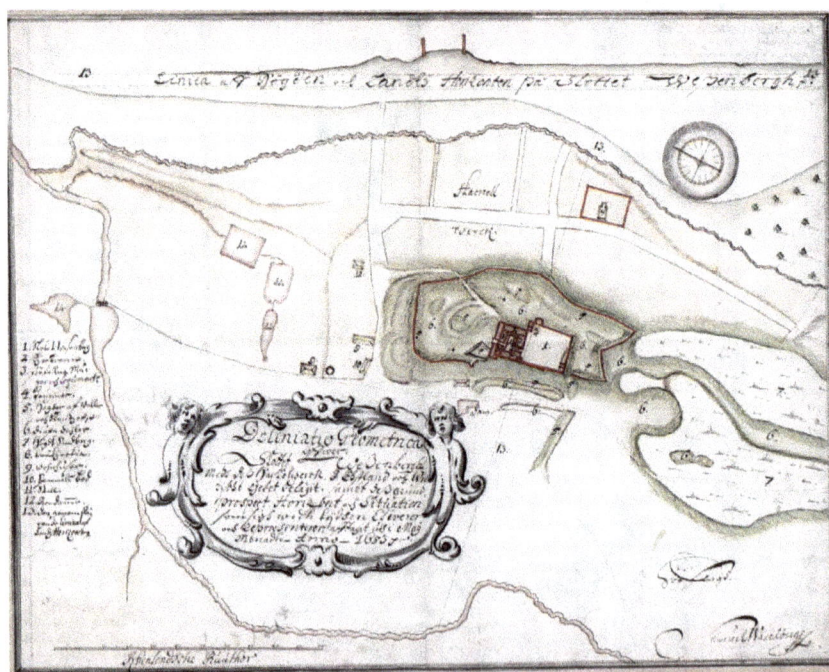

Figure 6.2: The castle of Wesenberg (Rakvere) had crucian carp ponds depicted on the map from 1683 (KA 0406.28.057.002).

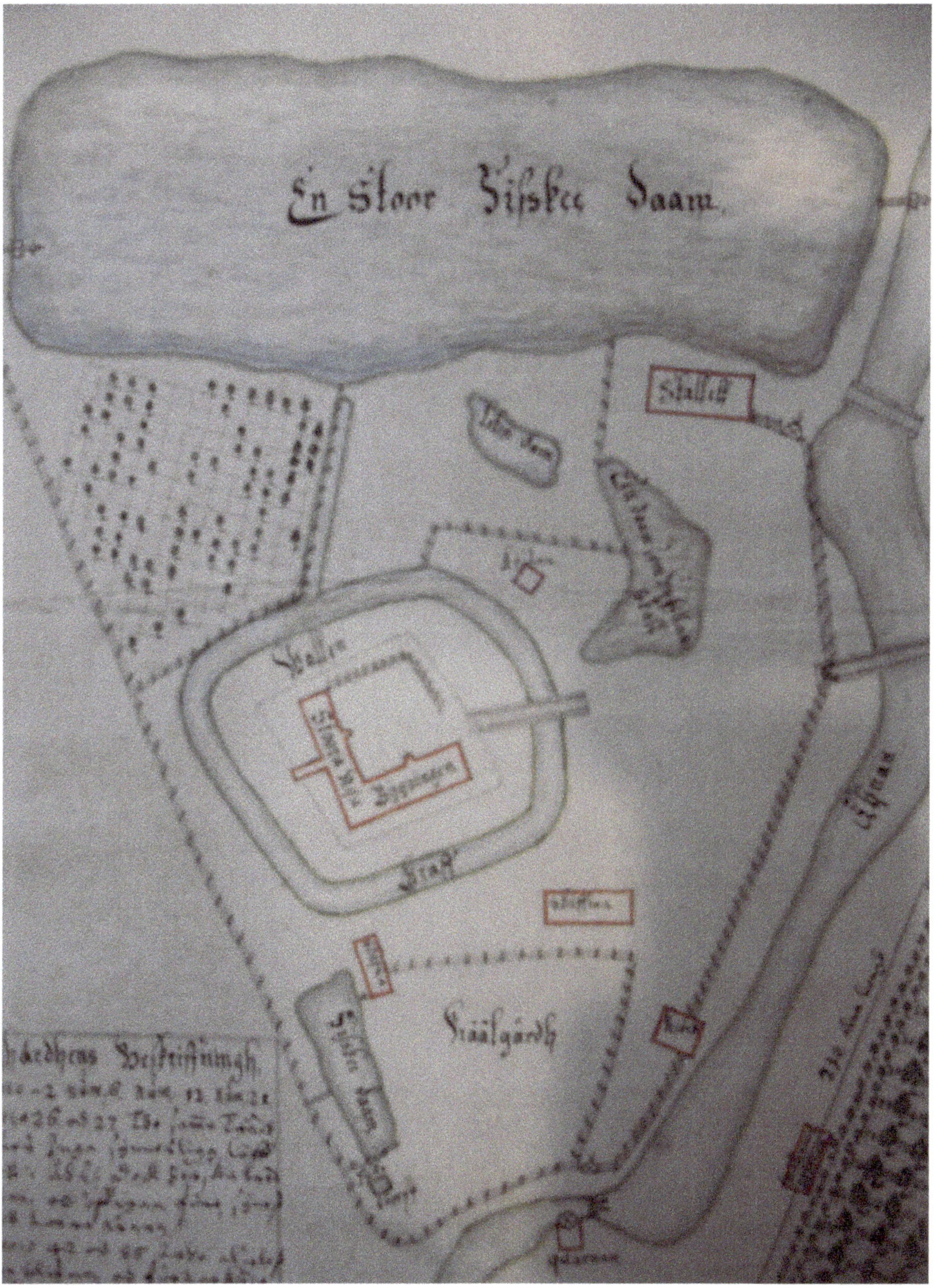

En Stoer Fiskee daam.

Figure 6.3: The surveyor Faber's layout plan (1649–54) of Marshall G. K. Horn's estate at Vainiži Wainsel and the Baroque garden (RA, 2025: 02). (The blue colour is fish ponds)

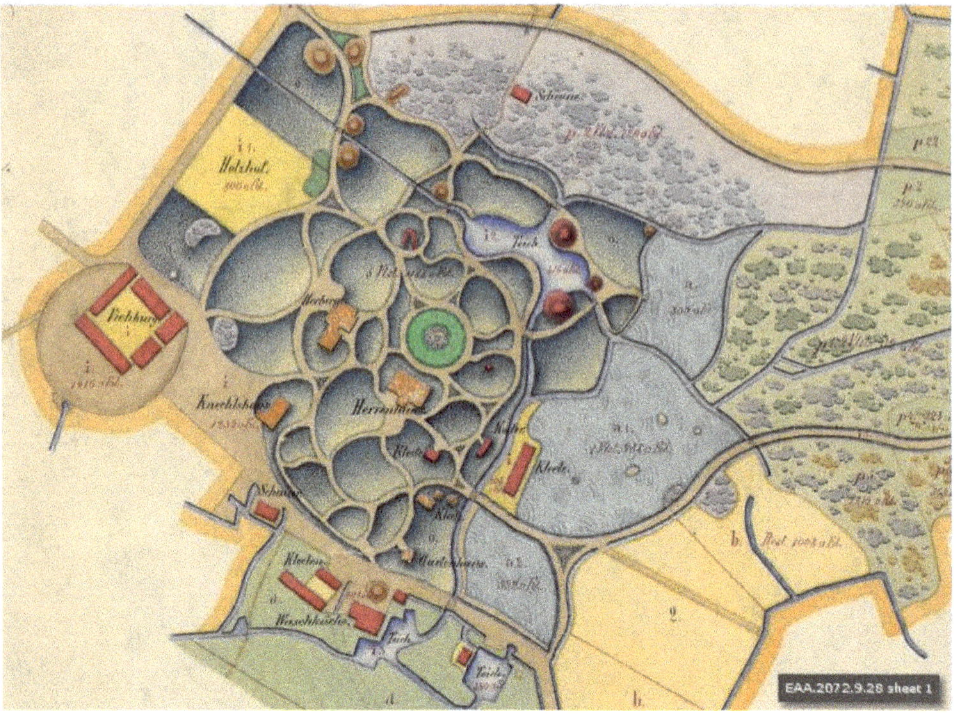

Figure 6.4: Pöllküll gehörigen Parke 1882 (EAA.2072.9.28)

CHAPTER 7

"The Increase of those Creatures that are Bred and Fed in the Water"[1]

Fishponds in England and Wales

James Bond

> *Pysgodlyn, cudduglyn cau,*
> *A fo rhaid i fwrw rhwydau;*
> *Amlaf lle, nid er ymliw,*
> *Penhwyaid a gwyniaid gwiw*

The lines above come from a Welsh verse by the court poet Iolo Goch (c.1320–1398) praising Owain Glyndŵr's castle at Sycharth. In truth Sycharth was an old-fashioned and rather basic stronghold; but the poet's imagination equips it with all the amenities of a grand English aristocratic residence, including orchard, vineyard, deer park, rabbit warren, mill, dovecote, and "a fishpond, enclosed, sheltered and well-stocked, into which nets may be cast to make catches of fine pike and *gwyniaid*". In modern Welsh *gwyniad* means salt-water whiting, but Iolo Goch is almost certainly referring to *Coregonus pennantii*, a member of the salmon family which occurs in Lake Bala and other Welsh lakes. Although Iolo Goch's vision cannot be interpreted uncritically as a literal description of the castle's setting, nevertheless earthworks there confirm that the fishpond was real enough (Hague and Warhurst 1966: 112).

The abundance of literary references to fishponds shows that their possession, along with mills, dovecotes and deer parks, was one of the privyleges of manorial landholders, a badge of rank as much as a practical utility. The Norman-Welsh cleric Gerald of Wales (c.1145–1223), extolling the delights of his childhood home at Manorbier, wrote that, just beneath the walls of the

---

[1] Izaak Walton, *The compleat angler*. 5th edition (1676): 1909 reprint. p. 23.

castle, "there is an excellent fishpond, well constructed and remarkable for its deep waters" (Thorpe 1978: 150). By the later middle ages enterprising men of lower status were also aspiring to the possession of fishponds. Geoffrey Chaucer (c.1340–1400), introducing his characters in the Prologue to the Canterbury Tales, describes the well-stocked household of the Franklin, a socially-ambitious freeholder who had "many a bream and many a luce in stewe" (Robinson 1950: 20).

There could be many motives for constructing fishponds: to fulfill fundamental subsistence needs; to contribute variety to a cereal, vegetable or meat-based diet; to reflect the status of an owner able to indulge in conspicuous consumption; to provide revenue from commercial sales; to promote aesthetic enjoyment as a feature of garden design; and to facilitate the enjoyment of angling as a pastime. Though these motives were rarely mutually exclusive, they varied in prominence according to place and time.

The investigation of fishponds

Ponds have served many different functions over the centuries, not all of them connected with fish. Assessing their date and purpose is rarely straightforward. Interpretations as fishponds may be supported by tradtion, by documentary records, or by physical characteristics and relationships; but scientific excavation of fishponds has rarely been seen as an archaeological priority, and even when undertaken does not always provide definitive answers. Identification of fishponds, therefore, continues to rely to a large extent upon circumstantial evidence.

Medieval records and antiquarian literature contain many scattered references to fishponds, and after the fifteenth century numerous treatises concerned with aspects of fishing and pond management appeared. However, there was little archaeological interest in the physical remains of medieval fishponds before the beginning of the twentieth century, when Hadrian Allcroft published sketch-plans and descriptions of several examples (Allcroft 1908: 465, 487–92). Interest then lapsed for another half-century, until publication of an influential volume discussing aerial photographs of medieval sites, several of which included fishponds (Beresford and St Joseph 1958: 54, 56–60, 68–9, 88–9; 1979: 53–4, 56–8, 67–9, 91–2).

Interest accelerated through the 1960s and 1970s. Charles Hickling (1962) explored connections between past and modern fishpond management. Fieldwork and documentary research by Brian Roberts in Warwickshire identified numerous fishpond sites, including no less than twenty examples within the 3,800-ha. parish of Tanworth-in-Arden, most of them in existence before 1350 (Roberts 1966). The Royal Commission on Historical Monuments in England published some fine archaeological surveys and interpretations of fishpond earthworks during their investigations in Northamptonshire, distinguishing seven main types on the basis of their physical form and setting (RCHME Northants. 1979: lvii–lix).

John Steane's study of Northamptonshire fishponds (Steane 1970) foreshadowed a further group of county surveys which appeared in a collection of essays edited by Mick Aston (Aston 1988). This volume was the first substantial compilation to focus specifically on medieval fisheries and fishponds in England, and it remains a valuable work of reference. Its contents were, however, assembled at a time when many inherited views were changing, and the editorial confesses that 'The conclusions to be drawn from this collection of papers are of some significance, since they were not anticipated before the papers were presented'. The most important revelations were that the effectiveness of the marine fish market had been greatly underestimated, that sea fish were consumed in much greater quantities by people at all levels of society, and that freshwater fish were a high-status luxury item rather than a staple element of the diet.

Further revisions to earlier views were discussed in several important contributions by Christopher Currie (1988, 1990, 1991). Since then, although fishponds have rarely been a primary focus of archaeological or historical attention, further work has continued to amplify our understanding.

The terminology of medieval fishing and fishponds

The investigation of medieval fishing practices from documentary sources is hampered by contemporary use of a wide range of terms, the precise meaning of which is often uncertain and may also change through time. It cannot be assumed that clerks recording matters relating to fisheries were personally familiar with fishery practices, so some confusion is inevitable. Though it is impossible to discuss in full here the origins and possible interpretations of all the Latin and Middle English terms encountered, common words for fishpond include *piscina, vivarium, stagnum, stank* and *stew*. Roberts (1986: 130, 132) and Currie (1990: 22–3, 43 n.3) recognised an important distinction between the *vivarium*, a relatively large pond in which fish were reared and fattened up from the natural feed available, and the *servatorium* or *cervorium*, a small square or rectangular pond, which served two temporary storage requirements: sorting fish when a vivarium was emptied; and keeping selected mature fish which were ready for eating. *Servatoria* for the latter purpose were located close to their owner's residence. Confusion has arisen through the Middle English vernacular term *stew* which, in the earlier middle ages, seems to be synonymous with *vivarium*, but later clearly came to mean a small holding-pond or tank in which live fish were kept until needed for the table (SOED: 2124). Records of repairs indicate that the Latin *caput* and vernacular *head* referred not to the upper end of the pond, but to the dam. The Latin term *baia*, Middle English *bay*, appears to relate to structural timber-framing within the dam (Moorhouse 1981: 714–5, 744; 1988: 476).

Identification of fish species from medieval and early modern records is equally problematic. Local dialect words, different terms for fish at advancing stages of their life-cycle and confusion between related species can all create difficulties. *Pickerel* become *pike* at a weight of 1.36 kg, still larger specimens being known as *luce*. Chub were also known as *chabin* or *chavender*, grayling as *umber*, minnows as *penk*. There is no generally-understood English vernacular term for Iolo Goch's *gwyniad*, though some early writers anglicised the spelling to *guiniad*; it has sometimes been confused with the *schelly* of the Cumbrian lakes (*Coregonus stigmaticus*), the Loch Lomond *powan* (*Coregonus clupeiodies*) and with the related *sewen* (*Salmo cambricus*) (Houghton 1879; 1984 edition: 201).

Edible freshwater fish in medieval Britain

The flavour of freshwater fish was affected by the character of the water from which they were taken, and the time of year, so recorded opinions on the edibility of different species varied considerably. Fashions also changed, and many species once enjoyed are now rarely eaten.

The family of Cyprinidae included the widest range of edible freshwater fish available in medieval Britain. The most esteemed was the common bream (*Abramis brama*), described in 1496 as "a noble fysshe and a deynteous" (Anonymous 1496); the price of a single bream was three or four times the daily wage of a skilled craftsman. Tench (*Tinca tinca*) were also widely kept. Bream and tench occur naturally in meres and sluggish rivers, and adapted well to artificial fishponds. Other native Cyprinidae eaten in the past include stone loach (*Barbatula barbatula*), gudgeon (*Gobio gobio*), rudd (*Scardinius erythrophthalmus*), roach (*Rutilus rutilus*) and dace (*Leuciscus leuciscus*). Bleak (*Alburnus alburnua*) and barbel (*Barbus barbus*) were less favoured, while chub (*Squalius cephalus*) were generally disdained as barely worth eating. Medieval sources record occasional consumption of *menuciae*, often translated as 'minnow' (*Phoxinus phoxinus*, smallest of the British Cyprinidae), but this term may refer loosely to any small fish or fry. The common carp (*Cyprinus carpio carpio*) was introduced into Britain only in the late middle ages.

Among other families of freshwater fish, the most valued representative of the *Esocidae* was the pike (*Esox lucius*), the natural habitat of which was in deep pools in slow-flowing rivers and weedy ponds. Although pike adapted well to artificial fishponds, it remained relatively scarce through the middle ages, and was very expensive. It also needed special care because of its omnivorous habits. The *Fleta*, a consolidation of two tracts produced around 1289 for the instruction of manorial bailiffs, recommends that "each prudent man see to it that his *vivaria, stagna, lacis, servoria* and fisheries of that sort are stocked with bream and perch, but not with pike, tench or eel, which strive to devour a profusion of fish" (Richardson and Sayles 1955: Lib.II, cap.73, 20). The family of Percidae included perch (*Perca fluviatalis*) and ruffe (*Gymnocephalus cernua*), while the burbot (*Lota lota*) belonged to the Lotidae.

Some edible species spent part of their life-cycle in the sea: the Anguillidae included several types of eel, including the common *Anguilla anguilla*, and the Clupidae the allis shad (*Alosa alosa*). Eels were caught in both rivers and ponds, remained relatively cheap, and were affordable by the less-wealthy. The Salmonidae included not only the migratory salmon (*Salmo salar*) and smelt (*Osmerus eperlanus*), but also freshwater brown trout (*Salmo trutta fario*), charr (*Salvelinus willoughbii*), grayling (*Thymallus thymallus*) and *gwyniad*. The Petromyzontidae included both the migratory sea lamprey (*Petromyzon marinus*) and the river lamprey (*Lampetra fluviatalis).*

Documentary records provide a more reliable guide to the types of fish kept in medieval ponds than archaeological remains. While wet sieving has greatly increased the recovery of fish bones and scales, the pond species identified may still be unrepresentative. Remains of larger fish generally survive better than smaller ones; and, since regular cleansing of a working pond normally removed dead fish, any remains recovered are likely to represent the last occupants of an abandoned pond, species most capable of surviving in weed-infested, stagnant and silting water (Chambers and Gray 1988: 126).

The origins of artificial fishponds in Britain

Even after fishponds became widespread in Britain, riverine and lacustrine fisheries continued to supply the bulk of the freshwater species consumed. However greatly they might be prized for the table or for sport, some migratory species, such as salmon, could only be obtained from estuaries and rivers; freshwater species requiring clear, cold, well-oxygenated water or clean gravel beds were also unsuited to ponds. However, reliance on natural resources always carried an element of uncertainty. Artificial fish–ponds provided a controlled environment for breeding and rearing young fish, for fattening bred fish or introduced river fish, and for storing fish ready for the table (Chambers and Gray 1988: 115).

Small ornamental fishponds have been identified within the courtyards of several Roman villas in south-eastern Britain, also several larger, more functional artificial ponds, including, at Shakenoak, a group of ponds possibly for breeding fish (Zeepvat 1988). However, these are unlikely to have survived beyond the early fifth century, and purpose-built fishponds were not seen again until the late eleventh century.

An intermediate stage could have been offered by artificial ponds powering mills, which were appearing in England by the late seventh century. Over 6,000 water-mills are recorded in the Domesday Survey of 1086, of which over 100 paid rents in eels. Eel renders from mills recur into the thirteenth century, when the Dorset mills of Sturminster Marshall and Marnhull each rendered 20 sticks (one stick being 25 eels) (Holt 1988: 3–14, 67, 96). An illustration in the fourteenth-century Luttrell Psalter shows two basketwork eel-traps set in a mill-pond. The flow of water through millponds also suited barbel, trout and grayling, in addition to bottom-feeding *cyprinidae* (McDonnell 1981: 14). Millers and fish-farmers required very different régimes of water storage and release, so compromises might be required to accommodate other fish; but on 28 January 1467/8 Sir John Howard transferred from the large pond in his park to his newly-constructed millpond 65 great bream, 66 little bream, 6 great carp, 240 little carp, 43 great tench, 20 small tench, 260 roach and 120 perch (Hudson 1841: 561–2).

Variations on the term *vivarium piscum*, probably implying a short-term store-pond for fish caught in rivers and intended for consumption, appear in the Domesday survey on two monastic properties (Chapter 1), also at Caversfield and on the holding of Osbern the Fisherman at Sharnbrook (*DB*, folios 148, 216v). The Norman aristocracy were responsible for the reintroduction of fishponds on a larger scale, and manorial ponds are frequently mentioned in charters, accounts, extents and court records from the late twelfth century onwards (McDonnell 1981: 1; Moorhouse 1981: 745) (Figure 7.1). By the thirteenth century professional fishermen were employed to manage ponds on royal, baronial and episcopal estates, supplying fish from them when needed (Hudson 1841: 16; Roberts 1986: 132; Steane 1988: 46).

Royal fishponds

Many royal castles, palaces, manor-houses and hunting-lodges were equipped with fishponds. Details of their management and exploitation have been summarised by Steane (1988). William the Conqueror's first fishpond was created in 1086–9 when he had the River Foss in York dammed to protect his newly-built castle. This created a large pool (*stagnum*) which destroyed two new mills and inundated some 50 hectares of arable land, meadows and gardens. It was stocked with bream and pike. The dam was swept away by floods in 1315, and its replacement thereafter required frequent repairs (RCHME York 1972: 60–1, 137–8; McDonnell 1981: 9–12). An itinerant royal household frequently absent in Normandy or Anjou had limited need for its own fishponds; nevertheless, examples had appeared on at least ten more crown properties in England before 1200, particularly in the north midlands. The pond at Stafford, first mentioned in 1157, formed part of the town defences, as at York.

During the thirteenth century the number of royal properties having fishponds tripled, the new acquisitions located mostly within the midlands, south and south-east. Henry III spent much more of his long reign (1216–1272) in England and was, therefore, more dependent upon home supplies. Direct demesne exploitation reached its zenith in this period. An expanding bureaucracy generated new classes of records, the Close Rolls and Liberate Rolls in particular providing increasing information on the management of royal resources. Fishponds at Marlborough Castle, Woodstock Palace, Havering Manor and the hunting-lodges of Brigstock, Clipstone, Feckenham, Kingscliffe, Silverstone and Woolmer remained in long-term use. Others were under royal ownership only intermittently or briefly. Specialist fishermen in royal employment travelled round England maintaining the ponds and supervising the netting and carriage of fish.

Annual feasts at the great religious festivals required large supplies of fish, often brought from a distance. Some royal ponds kept both pike and bream. Bream were the predominant product of the Feckenham ponds, whereas those at Woodstock produced mainly pike and eels. Pike were often transported live, but could also be salted or cooked in jelly; bream were often cooked in pastry or bread before despatch. When Henry III stayed at Kenilworth Castle he preferred to have fish sent from his pond at Stafford, over 60 km away, rather than exploit the lake and fishponds close by. Only a few of the royal ponds supplied the court on a regular basis: Marlborough provided fish on thirteen occasions between 1240 and 1272, usually when the court was at Windsor, Winchester, Woodstock or Clarendon. When the court was at Westminster it was usually supplied from the ponds of the Northamptonshire hunting-lodges, Brigstock, Kingscliffe and Silverstone. Gifts being a symbol of royal favour, the king made many grants of live fish from his ponds to favoured subjects. In 1231 the Dean of St Martin's, London, received 1,000 bream from Havering in a single consignment (Close R. 1231–4: 11). Marlborough supplied over 140 live bream to other landholders between the 1220s and 1260s, one consignment of six breeding bream being carried over 160km to Stamford (Close R. 1227–31: 470). The ponds at Feckenham, Woodstock, Silverstone and Brigstock also supplied breeding-stock for ponds elsewhere. By contrast, the ponds at Windsor, which probably lay near the moated lodge within the Great Park (Roberts 1997: 246–247), seem to have been used primarily as *servatoria*, storing fish for the royal tables.

The royal ponds were occasionally restocked by purchases: in 1250 the sheriff of Cambridgeshire was ordered to buy 3,000 pike for the king's stews at Havering, while in 1265 the constable of Windsor was required to buy 300 pike and 300 dace and roach to stock the Windsor Park stew (Liberate R, 1245–51: 273; 1260–7: 190). More commonly, live fish were transferred from other royal ponds: in 1247–51 400 bream were taken from Havering, 200 to stock the ponds at Kennington, 200 for the Windsor ponds (Close R. 1247–51: 399). Advantage was often taken of episcopal vacancies or wardships of secular manors to obtain fish from ponds temporarily in royal custody: in 1241 Henry III ordered 100 pike and bream to be taken from the Winchester episcopal ponds at Taunton (Liberate R. 1240–5: 87); in 1254 the bishop of Lincoln's pond at Banbury Castle provided 10 breeding bream for two royal servants; and in 1281 40 fat female bream, 20 other bream, 40 great pike and up to 400 other fish were transferred from the Bishop of Winchester's pond at Frensham to the royal stew in Windsor Park (Close R. 1253–4: 18; 1279–88: 79).

Episcopal fishponds

References to fishponds abound in the Pipe Rolls of the bishopric of Winchester between 1208–9 and 1399–1440 (Roberts 1986). Seven of the bishop's Hampshire manors included at least one pond, Marwell having four ponds in the early thirteenth century. More distant properties at Frensham, Brightwell, Harwell, West Wycombe, Witney and Taunton also had ponds. The surge of fishpond construction was almost certainly initiated after about 1150 by King Stephen's brother, Henry of Blois, bishop of Winchester, whose immense wealth enabled him to indulge in the latest fashions in high-status food and estate exploitation. Further ponds were added after Bishop Henry's death in 1171, the last recorded constructions at Bishops Waltham and Highclere completed before 1316. The ponds provided pike, bream, perch, and roach, which were rarely carried more than a day's journey from the place where they were caught, and were almost always eaten fresh.

Although the bishops of Winchester maintained over 160 ha. of ponds on their estates, it has been estimated that barely one tenth of their potential was exploited. In common with other seignurial ponds, they were treated as a personal luxury, the catches were reserved exclusively for consumption at feasts by the bishops and their close associates, and little attempt was made to maximise their yields. The only recorded sales result from the draining of the ponds at Alresford in 1252–3 and at Bishops Waltham in 1257–8, when eels were recovered in great numbers. Some were salted and the surplus sold. The quality of documentation declines after about 1350, but some of the ponds were still maintained into the early fifteenth century, though others had been leased out with their demesnes (Roberts 1986: 125–6; Currie 1991: 98).

A survey of the Bishop of Worcester's estates in 1299 notes fishponds on several properties. Four ponds adjoined the moat of Hartlebury Castle, valued at 100s every five years; another lay in Northwick, worth 50s every five years. The most complex group of ponds lay within the park at Alvechurch (Figure 7.2); here the fishery within the park and the 'fishery of the dug-out waters' (*piscarie aquarum infossatarum*) and the breeding fish retained to increase stock were estimated to be worth 25s every five years; in addition the pasture of the two islands in the fishery yielded 2s annually, while the fishery of the mill pond would yield 3s yearly if it was cleaned and stocked (Hollings 1934: 2, 7, 26; 1937: 190, 209; Aston 1970–2). The estimated valuations every five years carry implications about the management regime, discussed further below. Earthwork remains survive on each documented site.

In addition to their pond at Banbury, the bishops of Lincoln also had a moated palace in the deer-park at Stow with three large and two small fishponds, first documented in the 1180s (Everson 1991: 9; Everson et al. 1991: 184–5), and a well-preserved complex at Lyddington consisting of four rectangular ponds enclosed by a moat, with a deeper, larger pond just outside it (Hartley 1983: 26). The moats and fishponds around the bishops of Ely's palace at Somersham were designed to enhance its appearance in addition to their more obvious functions (Taylor 1989), and this is also likely to be the case on other episcopal sites.

Baronial and manorial fishponds

From the twelfth century onwards castles and manor-houses of the aristocracy and gentry were often surrounded by moats which, containing relatively stagnant water, were themselves suitable for bream and other Cyprinidae, eel and pike (McDonnell 1981: 14). Frequently there were also groups of fishponds of varying complexity nearby (Aberg 1978). Occasionally details of fishpond management appear in estate accounts, such as those of John Howard, Duke of Norfolk, from 1462 to 1471 (Hudson 1841: 560–565), and ponds are often noted in manorial extents and surveys. Fish-bones from excavations at Castle Acre in 1972–7 included an unusually high representation of freshwater fish, especially pike (over one-third of the total) and eel (Lawrance 1982: 287–9, 293; 1987: 297). Because fishponds ranked alongside deer-parks and rabbit-warrens as symbols of seignurial privilege, they too occasionally suffered damage and theft during times of peasant discontent (Dyer 1988: 35). In 1537 a local feud at Mangotsfield led to sixty people breaking down a fishpond, letting out the fish, stealing tench, bream and carp worth over twenty pounds and destroying fry to the value of twenty marks (Dennison and Iles 1985: 34).

Recent research has been less concerned with the practical function of seignurial fishponds than with their significance as a component of aesthetically designed landscapes, ornamenting both the approach to grand residences and the views from them. Large ponds or lakes are prominent in the settings of some castles, including Bodiam, Baconsthorpe, Kenilworth, Framlingham, Ravensworth, Clare, Stokesay, Ruthin and Caerphili (Taylor et al. 1990; Everson 1991: 9–12; Taylor 2000; Liddiard 2005: 45, 106, 114–5, 132–3; Creighton 2002: 75–80).

The construction of fishponds

Although natural lakes and meres were stocked with fish, the processes of breeding, rearing and redistribution could be regulated more effectively in artificial ponds. It was necessary to be able to control the water level in each pond and to empty it at will; to manage each pond independently of others in the system; and to redirect the flow of excess water. Careful site selection and sound construction of water channels, sluices and dams were essential (Roberts 1988).

Water fed into ponds directly from ground-water seepage or springs was unlikely to carry much pollution, but might be oxygen-deficient, and the colder water temperature through the summer could inhibit feeding-rates and growth. Spring-fed ponds remained relatively disease-free, unless diseased stock were accidentally introduced. Conversely, stream- and river-fed ponds received more silt and pollution, and were more vulnerable to disease, but also contained more water-borne food organisms. Acid water provided a hostile environment, but alkaline water encouraged the growth of zooplankton to supplement the food supply. Water quality influenced the selection of fish kept: bream, tench, perch and carp were relatively tolerant of low oxygen levels (Chambers and Gray 1988: 119–20).

John Taverner made an important distinction between fishponds on flat land, which might be surrounded on all sides by embankments constructed from spoil excavated from their beds, and required more careful surveying; and those made by constructing dams across narrow, steep-sided valleys (Taverner 1600: 2). Open meadowland sites might experience extensive shallow-water flooding, but ponds in constricted valleys were more vulnerable to the concentrated volume and accelerated force of floodwater, which occasionally threatened even large dams.

The creation of ponds demanded a range of practical skills. Digging out a simple stagnant-water storage-pond on flat clayland required merely muscle-power, spades, shovels and barrows. Retaining and controlling the level of water within a valley pond required knowledge of how to build a stable, watertight dam, how to puddle the bed with impermeable clay, and how to construct, maintain and operate sluicegates. More complex pond systems on level ground required even greater skills in surveying and constructing supply leats and drainage channels.

It is both curious and frustrating that so little contemporary documentation exists for the equipment and methods employed by medieval water engineers in Britain, despite ample visual evidence for the practical application of their skills. The first published accounts of techniques used in fishpond construction appear only in the middle of the sixteenth century.

A treatise on carp and pike ponds published in 1563 by Janus Dubravius, bishop of Olmütz (now Olomouc in Moravia) became known in Britain through an English translation in 1599. Describing the survey procedures required to determine water levels and dam heights, Dubravius (1599: 9–10) states that the instrument most widely used was the *dioptra*, a device described by classical writers, including Hero of Alexandria and Vitruvius; other levelling instruments were also described by Vitruvius. However, classical and renaissance writers tended to be more interested in instruments towards the top end of contemporary technology, which were not necessarily widely employed in practice. More basic devices, such as an A-frame with a crossbar providing the sight-line, levelled by means of a plumb-line suspended from the apex, were easily and cheaply made, and perfectly adequate for determining contours and laying out leats over short distances.

The techniques first described in the sixteenth and seventeenth centuries were undoubtledly evolved from centuries of experience by trial and error. The impressive works around the bishop of Worcester's palace at Alvechurch reflect the sophisticated achievement of medieval pond engineers. The palace itself was surrounded by a moat perched above the valley of the River Arrow, with steep slopes below it on two sides. To supply the moat a contour leat was constructed from the edge of the park, some 1,600 m up a tributary valley. The leat's point of diversion from the stream and its entry into the moat are not intervisible, so significant surveying skills were required. To avoid too steep a fall, the overflow from the moat was then conducted *up* the flank of the main valley until it encountered the river. The stream flowing down the tributary valley below the leat to the moat was itself diverted into a second contour leat at a lower level in order to form a large millpond and a complex of smaller fishponds with islands in the valley floor. The ponds existed before 1299 (Hollings 1937: 209–10; Aston 1970–2).

Over permeable subsoils, pond beds needed to be lined. A small fishpond made in the queen's garden at Rhuddlan Castle in 1282–3 was puddled with four cartloads of clay brought up from the nearby marsh (Brown et al., 1963: i, 324). Aston (1982: 276) has noted the common proximity of clay- or marl-pits to fishponds in Worcestershire.

Where dams were required their effectiveness was critical and their construction expensive. Later writers gave varied and sometimes conflicting advice on the advantages and disadvantages of forms of construction employing rammed earth, clay, turf, wooden piling, brushwood, timber framing and stone revetment. Dubravius (1599: 10) emphasized the importance of a good foundation using rammed earth, and Taverner (1600) recommended building up dams with successive later of fine earth through spring or autumn, watering each successive layer to bind and settle it. Dubravius and Taverner advised that the dam's width at its top should be the same as its height, and that its base should be three times wider. Broad earthen dams with shallow batters on either side were, indeed, commonly used. Dubravius, Markham and Walton all recommended strengthening the dam with rows of stakes or posts of elm, oak or ash to contain a core of brushwood or clay; Walton (1676; 1906 edition: 198–201) advised charring the piles in a fire before driving them into the ground to inhibit rotting.

As noted earlier, the term *bays*, used in relation to fishponds, indicates that some dams were reinforced by timber-framing. In 1265 oaks taken from Savernake Forest were carried to Marlborough Castle for various repairs, including new bays for the fishpond; in 1298 repairs to the bays of the Marlborough stewpond used ten more Savernake oaks (Close R. 1264–8: 71; 1296–1302: 167); further repairs to the same pond in 1271 consumed sixty oaks from Savernake and ten from neighbouring woods (Close R. 1268–72: 329, 353). In 1240 sixty oaks from Feckenham Forest were allowed for repairing the bays and stew by the king's house there (Close R. 1237–42: 229–30), and further oaks from the park and forest were used in subsequent repairs into the later thirteenth century. Repairs to the royal pond at Newport in Essex in 1231 used twenty oaks from the king's wood outside Hatfield Park and ten more from within the park (Close R. 1227–31: 531). A contract drawn up in 1377 for repairs to the Foss Pool dam in York required the old and weak timber-framing of a damaged section to be replaced with new timber-framed cross-braced bays, which were to be filled up with large stones and clay (Harvey 1975: 193–4).

No evidence has been found for any medieval dams in England constructed entirely of masonry. Dubravius (1599: 12–13) indicated that high costs precluded stone dams, though he appreciated their greater longevity. Nevertheless, stone revetting occasionally appears. In 1298–9 the long fishpond at Leeds Castle (Kent) was repaired with stone and lime (Brown et al. 1963: ii, 697). Lime was also used in repairing the bays of the king's fishponds in Woodstock Park in 1239 (Liberate R. 1226–40: 414), and surviving remains of a stone-revetted dam in the same park almost certainly belonged to a short-lived water-mill dismantled in 1334 (Bond and Tiller

1987: 34–37). Some fourteenth-century Yorkshire manorial accounts record purchases of stone in large quantities to be laid at the core of mill-dams and to repair breaches in them (Moorhouse 1981: 713).

The potential complexity of dam sequences can be illustrated by a site in the valley south of Wharram Percy church. Excavation here revealed a primary low clay dam, strengthened at the front with wattle hurdling, probably serving a late-Saxon horizontal-wheeled water-mill. This was enlarged in the early thirteenth century, probably for a new vertical-wheeled mill, which itself fell out of use after the 1250s. The early dams were then sealed by a much more massive dam of chalk rubble and earth, as the redundant millpond was converted to a fishpond. This may have belonged to the new parsonage-house built by the Augustinian canons of Haltemprice Priory following their acquisition of the advowson of Wharram Percy church in 1327. The pond continued to be maintained after the suppression of the priory and desertion of the village, and the dam was refaced with sandstone blocks in the eighteenth century (Beresford and Hurst 1990: 65–7, 103).

The regulation of water levels within each pond was achieved by a combination of supply leats, diversion channels and adjustable sluicegates. During construction of valley-bottom ponds the entire flow of the natural watercourse had to be diverted into a bypass leat along the side of the valley while the pond bed was prepared and the dam built. The side leat continued to perform an important function after completion of the pond, by diverting excess water. In 1342–3 a new bypass channel was dug around the Bishop of Winchester's ponds at Frensham to reduce the risk of sudden floods sweeping away the dams. At Harrington the main stream was diverted along the valley side, each of the three ponds could be filled independently from channels tapping springs on either flank of the valley, each pond was equipped with a lateral overflow leat, and any of the ponds could be emptied without affecting the others (Beresford and St Joseph 1979: 68–9; RCHME Northants. 1979: 77–8).

Preventing the level of water within a pond from rising too high could be achieved by a stone-lined spillway (McDonnell 1981: 32, 35). North (1714: 12) recommended setting one at either shoulder of the dam, protected by a latticed barrier to keep floating debris from blocking it. Some of the Winchester episcopal ponds seem to have had sophisticated penstocks or sluices set within the dams so that they could be emptied at regular intervals (Roberts 1986: 134–5). However, sluice-gates within dams were prone to leakage, causing erosion and, potentially, considerable damage, if not well maintained. In post-medieval ponds this problem was overcome by means of a vertical pipe set back from the dam with its top at the required water surface level, conducting the overflow into a sub-surface culvert passing beneath the dam to an outlet lower down-valley (Roberts 1988: 10–13). A plank-covered culvert of hollowed-out tree trunks beneath a seabank at Newton (Cambridgeshire), radio-carbon dated to the mid-thirteenth century (Taylor 1977), shows that this solution was not beyond the capacity of medieval technology. When the demesne ponds of Baddesley Clinton were emptied in 1443–8 a carpenter was employed in 'making a pipe' in one of the ponds (Roberts 1966: 123).

Dubravius (1599: 13) and North (1714) recommended oak for sluicegates. In 1252 timber from Wychwood Forest was used to repair the sluices of the stewponds in Woodstock Park and timber from Feckenham Forest for repairing the sluices there; soon after, 20 young oaks were required to repair the sluices of the great fishpond at Clipstone (Close R. 1251–53: 167–8; 1253–54: 79). Work on the Winchester episcopal pond at Alresford in 1252–4 required construction of at least nine new sluice-gates, equipped with winding-gear lubricated by grease. In 1299–1300 seven carpenters worked for five weeks on the sluice-gates around the dam of the great pond at Marwell (Roberts 1986: 135).

The size and form of medieval fishpond

Today fish are normally bred in small ponds where eggs can be fertilised, and fry, yearlings and two-year-old fish segregated and protected from natural predators (Roberts 1988: 12–15); once grown to maturity, they are transferred to larger lakes where fishing takes place. Medieval practice was different. Artificial breeding and specialised breeding-tanks were things of the future. Stocks were maintained by the seasonal produce of fry brought from rivers, and by transfers of breeding-stock between ponds. A basic distinction was made earlier between the larger *vivaria*, in which fish were allowed to breed and grow, and the smaller *servatoria*, holding-ponds where fish ready for eating could be stored live until needed for the table. However, some flexibility was possible, and rearing ponds could be converted to store ponds for mature fish in the autumn (Chambers and Gray 1988: 116). There were advantages in maintaining separate ponds for fish at different stages of growth and for mutually incompatible species (pike being especially notorious for their omnivorous habits).

In the middle ages the capacity of a pond to support fish was determined largely by the surface area of water absorbing oxygen from the air (Chambers and Gray 1988: 120). Ponds which remained in use over a long period are likely to have varied in extent, and their original size may now be difficult to determine. Some ponds of the Bishops of Winchester were conspicuously large: Frensham Great Pond occupied 40 ha., the Taunton *vivaria* 28 ha., Alresford Pond 24 ha. The larger ponds tended to be further upvalley, where small headstreams could more readily be diverted into bypass channels when they needed to be drained. From the evidence of sites where earthworks survive undisturbed by later land use, the majority of ponds ranged between 0.4 ha. and 1 ha.

Water depths of less than 0.6 m rendered the fish vulnerable to heron and other predators and, in hard winters, were more likely to freeze solid. Conversely, though some ponds in steep-sided valleys exceeded depths of 2m this gave no advantage, since lack of sunlight penetration excluded natural feed at greater depths.

The shape and size of medieval fishponds depends to some extent upon the terrain. Taverner's basic distinction between flat open sites and con-stricted valleys has been noted; but later classifications accommodate a greater variety of forms (RCHME Northants. 1979: lvii–lix). In constricted valley sites ponds tend to be long and narrow. Some were formed simply by constructing a dam across the valley, and these usually taper towards the upper end. Such ponds could be made deeper and more extensive by removing spoil from behind the dams and cutting away the valley flanks to make artificial scarps, a process which tended to create more rectangular shapes. Valley ponds include both single examples and chains of several ponds in line. Ponds could also be made by digging out spoil from a hillside and using it to create a dam on the downslope side, fed either by spring seepage or by leat from higher up the valley.

On open sites a much wider range of forms occurs. Many medieval manor-houses and isolated farmsteads were surrounded by moats, of no particular defensive strength, and documentary records show that many moats were themselves used to contain fish. Other moats enclosed orchards or served as garden features, and these also often doubled up as fishponds. Many moated dwellings had one or more separate fishponds nearby, sometimes in arrange-ments of considerable complexity (Aberg 1978). Groups of between seven and ten small rectangular ponds adjoining moated sites were recorded by Allcroft (1908: 489–491) at Rolleston and Sibthorpe. The Bishop of Lincoln's four ponds *within* a moat at Lyddington were noted earlier. A similar complex of five small ponds partly surrounded by two arms of a moat-like feature has been recorded at Kenn Court (Dennison and Iles 1985: 44, 46–7). A compact group of five small spring-fed ponds alongside the site of an unmoated manor-house at Somerton are documented in 1296 (Chambers 1977).

Groups of small ponds sometimes occur in association with a much larger pond. At Kenilworth Castle about fifteen small ponds occupy a compact rectangular area in the flat floor of a tributary valley just below the great mere, which itself fulfilled the threefold purpose of containing fish, reinforcing the castle defences and providing a spectacular visual amenity (Aston and Bond 1988: 420, 422–3). An unusual arrangement occurs at Braybrooke, where the moat of the manor-house lies on the southern side of a valley, with a complex series of earthworks on the slope to the west, including one large and two small flat-topped mounds defined by ditches on the uphill side and containing rectangular tanks; these are linked with two more ditched square flat-topped mounds without internal depressions, still further west. The entire group overlooks a much larger pond, nearly 300m in length, retained at the western end and along the northern flank by a massive dam up to 2 m high. The valley-bottom stream is deeply entrenched along the northern flank of the large pond at a lower level, and it is not easy to see how any of this complex was fed with water (RCHME Northants. 1979: 11–13).

Despite Currie's cautionary note, groups of small rectangular ponds in close proximity are often identified as ponds for breeding and rearing immature fish, segregating those of different size and age (Roberts 1988: 14–15). However, where they occur in proximity to a castle or manor-house, as at Kenn, Somerton and Braybrooke, they are more likely to serve as holding-ponds containing live fish ready for the table. Larger, elongated narrow ponds, which may be laid out in either linear or parallel arrangements, seem to have been designed for the use of trawl nets, and appear to date mostly from the late middle ages (Chambers and Gray 1988: 122).

Pond systems could be altered or extended. In Woodstock Park the earliest records mention only a single pond, but in 1242 the bailiff was ordered to repair the causeway between the two fish-stews, and in 1252 an order was issued for the making of a third stewpond in the king's garden (Liberate R. 1240–45: 164; 1251–60: 24–5). Individual ponds could be enlarged by raising their dams. The great mere at Kenilworth was retained by a massive dam at its east end, which also served as a causewayed entrance to the castle. Archaeological examination in 1965 showed that the original dam rose about 2.4 m above the valley floor, linking two natural spurs of higher ground. In the early thirteenth century the dam's height was raised by about 3.7 m, its length increased to 152 m and its width across the top to 15 m, which increased the mere's extent to over 40 ha. The dam was deliberately breached after the Civil War to drain the lake (Thompson 1965; Aston and Bond 1988: 420–3).

Ponds were inherently subject to silting, particularly at the entry-point of the feeder leat, and if this process went unchecked the pond would ultimately become choked up. Modern fishpond systems usually include separate small silt-traps enabling the bulk of the sediment carried in suspension by the incoming streams to settle out before the water was fed into the main ponds. Although settling-tanks were familiar by the mid-twelfth century in monastic water-supply systems, and Currie (1988: 282–3) has noted a silt-trap above the fishpond at St Cross Hospital by Winchester, silt-traps do not seem to have been employed routinely at fishponds, probably because the regular emptying of medieval ponds, discussed below, also provided opportunities for carting away surplus accumulations of sediment. In Sussex and Surrey the silt was itself seen as a valuable by-product of fishponds by the later middle ages, used for manuring the fields (Thirsk 1967: 167).

Allcroft's early surveys of fishponds at Langley Marish and Flamborough suggest deliberate intents to provide contrasting areas of shallow and deeper water (Allcroft 1908: 488–90), and a similar provision at Battenhall was noted in Chapter 2. Some fish preferred shallow areas for feeding, deeper water providing refuge from herons. This arrangement also assisted harvesting, the fish becoming concentrated in the deepest areas as the water level was lowered.

Even where ponds were designed primarily for fish, they could accommodate other uses. Osiers planted around the head of the king's fishpond at Kennington in 1247 (Liberate R. 1245–51: 108) probably provided the raw material for making basketwork traps. Reeds and rushes around the margins were strewn on floors and used for thatching. Ponds attracted wildfowl, which could also be eaten. Swans were kept on several of the Winchester episcopal fishponds (Roberts 1986: 135). Many ponds were equipped with shallower marginal embayments, perhaps to provide easier access for ducklings. Many also contain islands, of varying size (Aston 1982: 275–6). It is rarely known for certain whether these date from the pond's initial construction, or what their function was. Their most likely purpose was to generate reed-beds and provide shelter and nesting-places for swans, ducks and other wildfowl, but some may have been designed purely for ornamental effect.

Protection from casual poaching and stray animals was often provided by enclosing fishponds within a barrier. The great pond at Marlborough was enclosed by a hedge in 1239, and in 1256 a hedge was to be made around the king's stewpond in his garden at Woodstock (Liberate R. 1226–40: 415; 1251–60: 272). Moorhouse (1981: 744) has noted several examples in Yorkshire of fishponds enclosed by hedges, at Nostell Priory (1229), Calverley (c.1290), Pickering (1324–5) and Rothwell (1341). The fishpond made as an amenity in the queen's garden at Rhuddlan Castle was to be surrounded with seats (Brown et al. 1963: i, 324).

Enclosures around ponds could also contain buildings providing temporary or permanent domestic quarters for employed fishermen, stores for nets and other equipment, and fish-smoking houses (smoking freshwater fish was intended to enhance their flavour rather than to preserve them) (Moorhouse 1988: 477–9). Fish kept temporarily in shallow holding-ponds awaiting redistribution were especially vulnerable to poachers, and the Bishop of Winchester's ponds at Alresford and Frensham were provided with lodges to shelter watchmen night and day. The Alresford lodge was thatched in 1253–4 (Roberts 1986: 133). Cardinal Wolsey's London palace, York Place, had a fish-house supplied with piped water where fish were kept alive until needed for the table (Thurley 1993: 156).

The management of medieval fishponds

While there were advantages in making multiple use of medieval ponds, main-taining a satisfactory ecological balance was difficult. Artifical fishponds pro-vided a specialised environment intended to favour herbivorous and omni-vorous species tolerant of relatively warm, slow-moving and poorly-oxygenated water (Hoffmann 1994: 404). They were not, however, a stable environment, since silt accumulated and plants took root. Aquatic plants aerated the water and encouraged algae and other food elements, but fish sizes declined and yields were reduced if plant growth became excessive. Water-lilies, by limiting the extent of surface water able to absorb oxygen, were particularly damaging. Moreover, fish had many natural predators, including otters, herons, bitterns, kingfishers, swans, geese, water-rats and other fish. Ducks and frogs ate the spawn of tench and carp. Pike were voracious feeders, and in 1332 the bailiff of Cuxham was accounting for duckings that had been eaten by *lupus aquaticus* (Harvey 1965: 38). Like any other artificial environment, therefore, fishponds required careful management.

Because fishponds inevitably became silted, fouled and overgrown, they needed to be emptied, cleansed and restocked periodically. On well-managed estates this took place every five years. The Bishop of Winchester's pond at Bishop's Waltham was emptied and restocked in 1247–8, 1252–3 and 1257–8, while income from the Bishop of Worcester's ponds was calculated on a quinquennial basis in the late thirteenth century (Roberts 1986: 132; Hollings 1934: 2, 7, 26; Hollings 1937: 190, 209).

Although water level might be lowered by opening sluice-gates and the task could be completed by baling, the usual way of emptying a pond, commonly recorded from the thirteenth century, was simply to break down some of the bays of the dam (McDonnell 1981: 35–6). This remained a regular practice into the eighteenth century, though North (1714: 15–16) disapproved of it, since repairs to the breach were never entirely secure and could cause leakage.

Fish from emptied ponds were transferred to holding-tanks, to separate those ready for consumption from those to be returned to other ponds for further growth. When the royal pond at Marlborough was emptied in 1270, the eels from it were salted and taken to Windsor and the pike taken to the king at Clarendon and to the queen at Guildford (Liberate R. 1267–72: 115). Eels, readily available from rivers, streams and mill-races, seem to have been taken from the Bishop of Winchester's ponds only when they were drained (Roberts 1986: 132).

Emptying a pond enabled silt to be removed. Mud accumulated to a depth of 1.5 m was dug out of the East Meon pond in 1231–2. In 1244–6 ten men spent 40 days clearing the same pond, using stretchers and 12 wheelbarrows to carry the mud away (Roberts 1986: 133). Between 1443 and 1448 the three demesne fishponds at Baddesley Clinton were repaired before restocking, the most costly operation being the removal of mud deposits between 1.2 m and 1.8 m deep (Roberts 1966: 122–3).

Taverner (1600: 10) advised leaving an emptied pond dry and exposed to the air for a season, and cultivating its bed for a cereal crop before reflooding and restocking it with fish. This practice is documented 350 years earlier on the Winchester episcopal estates: the Bishop's Waltham pond was dug with spades and planted with barley in 1257–8, and 60 acres of barley were sown on the Alresford pond in 1252–3 (Roberts 1986: 133). A number of former ponds contain traces of ridge-and-furrow indicating former ploughing, and in the large ponds at Braybrooke and Wormleighton this is confined to the beds and does not override the banks: ploughing therefore occurred while the ponds were still actively managed and not after their abandonment.

During the thirteenth century favoured members of the court and men in royal service were often provided with gifts of live fish from the king for restocking their ponds (Steane 1988: 46). Alternatively, new stock could be purchased from professional fishermen or selectively transferred from other ponds within the estate. Restocking ponds could be a costly and laborious operation, sometimes involving transport of live fish over considerable distances. In 1231–2 live bream from the Bishop of Winchester's pond at Taunton were carried in canvas-lined water-filled barrels to Winchester, a distance of 160 km, taking 15 days and costing £8 3s 9d; 30 men were then hired to carry the bream a further 13 km to the pond at Bishop's Waltham. In 1251–2 19 men were hired to transport 50 bream in casks from the king's pond at Woolmer to the bishop's pond at Frensham (Roberts 1986: 130–1). In 1443–8 two 45-litre barrels with lids were purchased to enable the demesne ponds at Baddesley Clinton to be restocked with bream, tench and roach from places up to 48km away (Roberts 1966: 123–4). Bream and tench could live for some time out of water, and over shorter distances might be transported packed in wet grass or straw. Though restocking was often carried out in autumn, winter was preferable since the fish suffered less damage in colder conditions. The Winchester Pipe Rolls record over the winter of 1254–5, the restocking of Alresford Pond after draining: 115 pike, 229 perch, 603 bream and 1,072 roach were transferred from four other ponds on the bishop's estate and from the king's pond at Woolmer (Roberts 1986: 131).

Wattle hurdles were commonly placed at the entries and exits of medieval ponds, permitting water to flow in from the feeder streams and out through a sluice or spillway, while at the same time preventing fish from escaping upstream or downstream. A fence of thirty-three hurdles was set across two feeders into the Alresford pond between 1254 and 1257, and hurdles set within a timber framework for support against the strong current were used to retain the fish in the pond at Marwell in 1283–4 when the sluices were opened to empty it (Roberts 1986: 133).

Fishing methods in the middle ages

Breaching dams, by reducing the water volume to shallows where the fish could easily be netted, was the most efficient way of taking a large catch. However, during the intervals between emptying, fishing by other means took place whenever the need arose. The bishop of Winchester's ponds were usually fished through the winter, especially for Christmas, Easter and Lent (Roberts, 1986: 131). Steane and Foreman (1988) have reviewed medieval fishing methods and equipment, including spears and harpoons, wickerwork traps, nets, and rods and lines. All of these could be used in fishponds, though some were more generally employed in river and estuarine fisheries.

Routine fishing of ponds usually employed nets. Medieval records rarely distinguish between different types of nets, although repair of 'stake-nets' (*haias vivarii*) at Brigstock in 1263 required 4 oaks (Close R. 1261–4: 224). Bulky nets were difficult to carry over long distances, and the Crown records occasionally record orders for royal fishermen sent to distant ponds to be provided with nets on the spot; the Abbot of Waverley provided nets in 1263 (Liberate R. 1245–51: 82–3; Close R. 1261–4: 228). An account book for 1632–6 kept by the Duke of Suffolk's estate steward records the cleaning-out of the Lulworth Castle fishponds at a cost of £9 4s 8d and the purchase of a 'trammell nett' (a long, narrow fishing-net held vertically in the water by floats and sinkers, consisting of two walls of large-meshed netting, between which a narrow-meshed net was loosely hung) for catching the fish (Bettey 1993: 79). Some ponds were commonly fished from boats. On the larger Winchester episcopal ponds fish were trapped by attaching one end of a long seine-net to the shore, then taking the other end out in a boat around a wide circle. Between four and eight men were needed to haul in the net. The Winchester Pipe Rolls also record the building and repair of boats and a boathouse at Bishop's Waltham, the construction of a quay at Frensham Great Pond in 1282–3 and the overland carriage of both nets and boats from one episcopal pond to another (Roberts 1986: 131). Net-weights, commonly associated with river-fishing, have been found in or near several fishponds in the Thames valley (Thomas 1981).

Angling with a rod, line and baited hook was a far less efficient means of catching fish, but it required understanding, patience and skill. Manuscript illustrations suggest that angling was becoming a popular recreation by the thirteenth century. The earliest English text to promote fishing as a sport, entitled '*The Treatyse of Fysshinge with an Angle*', was inserted by Wynkyn de Worde in 1496 into his second printing of *The Boke of St Albans*, a manual of instruction for gentlemen on hawking, hunting and heraldry first published ten years earlier. The attribution of this text to Juliana Berners, reputed Prioress of Sopwell, has long been discredited, and its true authorship remains unknown, though parts were derived from earlier sources, including an English manuscript dating from after 1406 now in the Beinecke Library at Yale, and still earlier continental treatises from the fourteenth century (Van Siclen 1880; McDonald 1963; Braekman 1980; Hoffmann 1985). The *Treatyse* provides instructions on making rods, lines of various colours, hooks, lead sinkers and cork floats, and on the handling of the equipment. It advises on the seasons, times and best places to fish, in both pools and rivers, and how to select and prepare bait for different fish at different seasons. Specific instructtions are given for catching salmon, trout, grayling, barbel, chub, bream, tench, perch, roach, dace, bleak, ruff, flounder, gudgeon, minnow, eel, pike and carp. It was known to many later writers such as Mascall, Barker and Walton, who borrowed information from it in their own work.

Freshwater fish production in the later middle ages

The fourteenth century witnessed a significant decline in the seignurial monopoly over fishponds and the practice of royal, aristocratic and ecclesiastical households supplying their needs from their own demesnes. Correspondingly, increasing demands for freshwater fish on the open market expanded commercial production considerably.

Declining use of the royal fishponds coincided with a reduction in the number of residences maintained by the crown, as government activities became increasingly centralised in London. More remote royal houses, with their ponds, fell into neglect. In 1428 the fishpond at Lyndhurst was reported to be full of mud and grass and the dam so broken that it no longer retained water (Patent R. 1422–9: 460). This was only partly compensated by the acquisition of new manors within a day's ride of the capital, such as Sheen, where Edward III had a new fishpond made and an old one cleaned out in 1358, and Isleworth, where he spent large sums upon the "stank of Baber" after 1369 (Brown et al. 1963: ii, 964, 995).

The long-standing notion of fishponds as a badge of status initially made many lords reluctant to relinquish them on long-term leases. When Sir John de Bishopton leased fishing rights in his ponds at Lapworth to two local fishermen over a six-week period spanning Lent in 1329, he reserved part of the catch to himself, including one pike or a gross of pickerel, one bream, one large eel, four *menes* ('minnows') and a dozen perch or roach daily for his own sustenance; also four brace of pike, four brace of bream and two dozen brace of perch and roach, half of which were to be retained for restocking the pond; and all the smaller pickerell, bream, tench, perch and roach, each of specified length, also kept for restocking. The lessees were taking a considerable financial risk, and clearly did so in the expectation of large catches (Roberts 1966: 124–5).

However, leasing-out of demesnes increasingly provided new opportunities for those lower down the social scale. Moorhouse (1981: 744) notes men of middling rank acquiring fishponds in Yorkshire by the early fourteenth century, while in Warwickshire many ponds were falling into the hands of the lesser gentry, free tenants and even wealthier peasants (Roberts 1966: 123–6). Much of the increasing quantity of freshwater fish available for sale during the first half of the fourteenth century was supplied by manorial lessees.

The largest market for fish was in the biggest towns. The royal provisioners found it more expedient to purchase from London fishmongers rather than to continue relying upon supplies from distant estates. In Southwark complaints about blocked ditches, flooding and the condition of riverside wharves in the 1360s name at least five professional fishermen who had extensive yards and ponds for holding and fattening fish destined for the market. This was clearly a substantial commercial business (Currie 1991: 100).

The rise in such businesses was accompanied by increasing efficiency in production. The principal reason for the low yields of medieval fishponds had been that fish foraged only on the natural resources of the pond. No efforts to increase stocks by supplementary feeding are recorded before the 1320s, but by the sixteenth century this had become common practice (Taverner 1600).

Another significant single innovation of the later middle ages was the intro-duction of the common carp (*Cyprinus carpio*) to Britain. It was tradetionally believed that carp had been introduced into Western Europe from China by way of Cyprus, hence their Latin name. However, the presence of indigenous carp in the Danube since the mesolithic period was established by Balon (1974) and confirmed by later investigators. Carp had been bred successfully in ponds in central Europe, France and Germany since the thirteenth century. Leonard Mascall, kitchen-clerk to Archbishop Parker of Canterbury, claimed to have brought the first carp into England at Plumstead (Mascall 1590), but records are known from a much earlier date. Carp mentioned in the royal kitchen accounts in the fourteenth century were probably imported into London from the Netherlands (McDonnell 1981: 1, 38 n.1; Currie 1991: 102). However, the author of the first printed English *Treatyse* on angling, derived from early fifteenth-century sources, shows first-hand familiarity with carp, describing it as 'a deynteous fysshe, but there ben but fewe in Englonde' (Anonymous 1496: Van Siclen 1880: 75).

The earliest unambiguous evidence for carp in English ponds comes from the accounts of Sir John Howard, Duke of Norfolk. Between 1462 and 1472 the Duke's steward placed some 800 carp, of which 142 were described as 'great carp' into seven of his fishponds. During that period 22 carp were har-vested after individual ponds had been emptied and 500 small carp returned to one pond to grow to maturity. On 27 September 1465 the Duke drained the largest pond in his park and reflooded it the same day, putting back 500 small carp and 40 small bream, and distributing 96 carp to five of his neighbours for stocking their ponds (Hudson 1841: 560–4).

Although carp did not breed particularly prolifically in Britain, they were hardy, able to survive cold winters and live transport; they tolerated cloudy water, low oxygen levels and some degree of pollution; and they thrived in the depths of small still-water ponds, out of reach of herons. Even more importantly, they required relatively little attention and grew more rapidly than most other freshwater species, so were particularly attractive to commercial producers. They remained relatively expensive: in 1538 five carp awaiting delivery to one of the royal ponds, valued at 5 crowns, were stolen from the store pond of a Suffolk supplier (L.& P. Henry VIII, 13.ii: 246). Nevertheless, their consumption at aristocratic tables was increasing and they were on their way to becoming the most popular fish for keeping and breeding in ponds, remaining so until the early nineteenth century (Currie 1991: 102–3).

From the reformation to the civil war

Although the Reformation ended religious dietary restrictions, economic and political pressures continued to encourage fish consumption. Meat remained expensive. Acts of Parliament in 1548 and 1563 re-established Saturday and Wednesday fish-days, partly to ensure a reserve of marine fishermen for wartime service at sea. Further proclamations against meat-eating in Lent and other fast days were issued under James I (1603–25) (Wilson 1976: 44–5).

Despite the plundering of some monastic fishponds immediately after the Dissolution, others continued to be exploited and expanded by subsequent owners (Chapter 1). New ponds were also created. Even if the prestige of freshwater fish had become reduced through their wider availability, high prices continued to ensure that only the wealthy could afford them. Throughout the later sixteenth and seventeenth centuries the choicest varieties continued to be stocked in ponds by many members of the country aristocracy and gentry for their own tables. Commercial production also continued to be important, at least for a time. However, ponds were increasingly becoming employed as an ornamental component of formal gardens, with fish production relegated to a secondary function (Currie 1990: 41–3).

While relatively few fishponds have been securely dated by archaeological methods, the juxtaposition of some examples with other landscape features can provide evidence of post-medieval origin. At Wormleighton the earthworks of four small ponds with a large rectangular pond above lie directly over the abandoned street of a village which was depopulated in the late fifteenth century; they were probably created during estate improvements in the sixteenth century (Aston and Bond 1988: 428).

Although three long parallel rectangular ponds adjoining a moated site near Thame resemble medieval examples, neither the documentary record nor evidence from partial excavation in 1973 have supported a medieval origin. Elm timbers, probably from a sluice-gate, were recovered, but no fish remains, since the ponds had been cleaned of silts. The moat, first recorded on an 1823 map, was occupied by a garden in the nineteenth century (Chambers 1975). Earthworks of an embanked rectangular pond at Cogges coincide with the location of a fishpond documented in 1228 and 1242. However, excavation of silts at its eastern end in 1984–5 produced a preponderance of terrestrial molluscs favouring damp shady conditions. The visible earthworks, therefore, relate to a short-lived pond which had become overgrown with scrub after its abandonment. Traces of an earlier, partly natural pond, extending beneath and beyond the western part of the visible earthworks, must be the pond recorded in the thirteenth century (Bond and Chambers 1988: 357; Chambers and Gray 1988: 124–125).

Many other palaces and manor-houses had gardens including rectilinear ornamental fishponds associated with various combinations of moats, islands, mounts and walks and, occasionally, larger lakes, which can be dated by documentary evidence. New ponds and gardens were made at Collyweston for Lady Margaret Beaufort in 1502–3 (RCHME Northants. 1975: 30–31) and at Woodham Walter Hall near Chelmsford for the Earls of Sussex (Everson 1991: 13–16). The damming of a stream to make the Great Pool alongside Raglan Castle was probably undertaken in the 1550s as part of the garden improvements designed for the 3rd Earl of Worcester, and a smaller formal water garden was added at its upper end in the 1590s by the 4th Earl (Whittle 1989). A square moated area with round prospect mounts at two corners formed part of the uncompleted gardens of Lyveden New Bield, made for Sir Thomas Tresham in the 1590s; contemporary correspondence records that its waters were intended to provide fish (Brown and Taylor 1973: 159).

Innovative geometrical designs characterize some ponds of the early seventeenth century. In 1612 John Harborne, a wealthy London merchant, bought a country estate at Tackley. Having built a new residence there, he then laid out a gated walk leading to water gardens some 400m to the east, which included two triangular ponds with arrow-shaped peninsulae (imitating angled bastions employed in contemporary artillery defences), a single square moat with a square island, walks between and around the ponds, and a viewing terrace above. As it stands this layout is obviously incomplete, and a second square pond must have been intended, but Harborne was never able to acquire the critical piece of land. A plan of the complete layout, including the missing pond, was inserted into later editions of several of Gervase Markham's treatises on farming and country living by Markham's publisher, Roger Jackson (Markham 1613, 1623, 1631). Jackson and Harborne were close friends and fellow-anglers, and the plan's caption denotes the primary purpose of the ponds, despite their ornamental form, as 'for the increase and store of fish' (Whittle and Taylor 1994).

Other ponds and lakes associated with early seventeenth-century formal gardens have been recorded at Kettleby near Brigg (Everson et al. 1991: 70–71), and at Chipping Campden (Everson 1989; Everson 1991: 12–13). Two new ponds and a mount were constructed at the royal palace of Theobalds in 1625–6, and the moat around the orchard there was 'reasonablie well stored with fish' in 1650 (Andrews 1993: 133, 144). In other cases, such as Sir Francis Bacon's elaborate water gardens at Gorhambury, begun in 1608, contemporary accounts provide no explicit mention of stocking with fish (Henderson 1992), yet it seems unlikely that their potential would have remained unexploited. John Norden's description of the ponds in Osterley Park emphasized their multiple use, not only for fish, swans and other water-fowl, but also to power paper-mills, oil-mills and corn-mills (Henderson 1992: 118).

One mid-sixteenth-century English medic, Andrew Boorde, proclaimed sea fish superior to river fish, and disapproved of eating fish from stagnant ponds or moats (Furnivall 1870: 268). In contrast to the continuing popularity of sea-fish, freshwater fish consumption does seem to have declined through the later sixteenth century, coarser varieties in particular falling out of favour. The Foss pond fishery at York had been granted away by the Crown in 1545 and, although it continued to supply fish for a time, its size and importance were reduced by the late seventeenth century (RCHME York, 1972: 61, 138).

Yet, contrary to this trend, a steady stream of texts containing advice on angling (Mascall 1590; Dennys 1613; Markham 1608, 1611, 1613, 1635; Barker 1653; Walton 1653) and manuals on the construction and management of fishponds (Taverner 1600; Markham 1613, 1614, 1623, 1631) continued to reflect and encourage these interests among the gentry. Gervase Markham advised that converting well-watered marshy land to fishponds could provide profit for smaller farmers, recommending two ponds covering 3.2 ha. or three ponds covering 6 ha., no deeper than 2–2.7 m, with at least four stew-ponds of about 10 x 15 m. This explosion of literature has suggested to some that the late sixteenth century was a time of innovation; but much of it is derivative, merely endorsing practices developed and employed over several previous centuries. Nevertheless, some improvements are suggested, to promote more intensive stocking of ponds. John Taverner advised fishing-out for stock-taking, sorting and redistribution every autumn; regular draining and cleansing of the ponds, leaving them dry through the winter and following summer so that exposure to sun and air could sweeten the soil and restore fertility; allowing grazing sheep and cattle to manure the area to encourage flies and other insects which would provide food for the fish once the pond was reflooded; weeding out rank growth to improve the quality of grazing; and cutting drainage trenches to dry out the bed so that it could then be planted up with summer corn. More importantly, he promoted the artificial feeding of pond fish, advising that, without this measure, an acre (0.4 ha.) of flooded ground would keep little more than 300–400 carp and other fish, whereas supplementary feeding with cheap grain could increase the numbers to three or four thousand (Taverner 1600: 12, 14, 16–17).

One reason put forward by Taverner (1600: 22) for advocating the keeping of carp was his concern that fish-keeping was declining in popularity. Incidental records note the presence of carp on many estates. A mid-sixteenth-century survey of Cornbury records three fishponds in the park stocked with bream, carp and eels (Bond and Chambers 1988: 366). In 1612 the Earl of Rutland's household treasurer purchased from a Lincolnshire fisherman quantities of pike, bream, tench and carp, including large fish for immediate availability and small fish for maturing in the stewponds of Belvoir Castle; the carp were still relatively expensive (Davis 1966: 128). Walton (1676; 1906 edition: 199–201) advised that carp did not do well in old ponds full of mud and weeds; they preferred sheltered ponds warmed by the sun, not too deep, with willows and grass on their sides and stony or sandy beds; they also thrived best when no other species were put in with them, since other fish devoured their spawn.

The growing interest in angling as a recreation and the accommodation of fishponds into rural gardens may reflect a peculiarly English idealisation of country life, perhaps derived ultimately from Virgil's portrayal of rustic contentment in his *Georgics*. William Lawson (1617) recommended placing orchards alongside streams or rivers to provide a pleasant spot for angling for trout, eel or other fish, or a moat sufficiently large to fish with nets from a rowing boat. Izaac Walton dedicated the 1653 edition of *The Compleat Angler* to his friend and fellow-angler John Offley, whose garden at Old Madeley Manor contained an elaborate set of ponds (Walton 1906 edition: 2–3). There was a vogue for building fishing pavilions between about 1570 and 1640 (Whittle and Taylor 1994: 52–53).

The Civil War and Commonwealth (1642–1660) disrupted the management of many country estates, and many fishponds fell into disuse. Although Raglan Castle's ponds still feature on a map of 1652, they had been abandoned following the ruin of the castle after the Civil War siege. In 1674 it was recalled that there had been 'a fishpond of many acres of land', with 'divers artificial islands and walks'; but after the castle's surrender in 1646 local men had breached the dam and drained the great fishpond, 'where they had store of very great carps, and other large fish' (Whittle 1989).

The late seventeenth and early-eighteenth centuries

During the Commonwealth fish-days were condemned as a Popish custom and abolished. Attempts to re-establish them by statute following the Restoration of the monarchy in 1660 failed. Nevertheless, many still followed the old customs voluntarily, and one mid-eighteenth-century cookery book was still providing fast-day recipes for fritters, puddings and pies based upon fish, eggs and vegetables (Glasse 1747).

Renewed continental influences upon garden design after 1660 introduced fashions for more rigidly geometrical ponds and straight canals, often with fountains (James 1712; Switzer 1734). Many medieval fishponds were converted. At Dyrham Park a chain of five valley-bottom ponds depicted on a 1689 estate map had been altered before 1704 and incorporated into a magnificent water garden, with additional cascades, canal and fountains (Dennison and Iles 1985: 34, 41). However, low-lying situations were increasingly regarded as unhealthy, and new garden ponds were often above the valley floor, supplied by pipes of bored-out oak, elm or alder, the joints sealed with pitch and clamped in iron, or by more expensive brick-lined conduits. Lead pipes were employed to work fountains, and iron pipes came into use during the eighteenth century. Stephen Switzer (1718: 307; 1734: 131–2) gave precise instructions on constructing dams of rammed chalk, cambering banks with wood, brick or stone revetments to a slope of 1 in 3 and puddling pond beds with clay topped by a layer of rammed chalk and lime. However, Switzer and other contemporary garden writers hardly mention fish. Many grand formal water features were simply too large to serve a secondary function as holding-ponds; and their high maintenance costs ultimately made them unsustainable (Currie 1990: 24–30, 35–36).

Yet many treatises on practical husbandry published during the later seventeenth and early eighteenth centuries continued to include advice on fishpond construction and management. Worlidge (1669), Blagrave (1685), Mortimer (1707), Bradley (1721) and Hale (1756) deal mainly with the larger breeding-ponds. Roger North (1714) expressed regret at the neglect of fishponds in more remote parts of the country, a consequence of landowners letting their estates to tenants; he suggested that a chain of ponds, in which the top of the lower pond reached to the base of the dam of the upper, 'may well be very beautiful as well as profitable'; and recommended that holding-ponds should ideally be 10 m wide and 15 m long, with sloping sides and an inclined bed to facilitate drawing with nets.

Izaak Walton's *The Compleat Angler, or the Contemplative Man's Recreation*, first published during the Commonwealth in 1653, enjoyed significant success, being revised and enlarged through four subsequent editions between 1655 and 1676. It influenced other angling texts (Venables 1662; Gilbert 1676). After a brief wane in popularity, a new edition issued in 1750 restored its reputation, and thereafter it achieved the status of a minor classic. Significant improvements in angling equipment during the mid-seventeenth century were described by Thomas Barker (1653, 1667) and Robert Venables (1662). These included the introduction of a wire ring at the top of the rod to accommodate a running line for casting and playing a hooked fish; the addition of a reel for taking up the increasing lengths of line; experiments with different materials for the line; improved methods of fish-hook manufacture; and the introduction of the landing-hook or gaff.

Izaak Walton relayed from Lebault's French text advice on ways of improving the numbers and quality of fish in ponds: the ponds should be large and fed with running water; they should have variable depth and beds, carp preferring gravelly bottoms, tench and eel preferring mud; willows or osiers should be planted on the margins, or bavins cast into the pond near the side to provide places for spawning and to protect the spawn and young fry from ducks, frogs and vermin; overhanging banks and tree roots provided retreats from predators and shelter in conditions of extreme heat or cold, but too many trees nearby, depositing an excess of leaves in the water, were harmful. Each pond should be cleansed once every three to four years, then allowed to dry out for six to twelve months in order to kill off weeds such as water-lilies and bullrush and to encourage the growth of grass; or alternatively, to sow oats over the dry bed. Supplementary feeding was also recommended (Walton 1676; 1906 edition: 198–200). Roger North was still promoting the culinary prestige of freshwater pond fish, recommending construction and management methods derived from medieval practices. North valued carp above all other fish, stating that they could attain 46 cm (about 1.8 kg, the usual size for commercial sale) within five years, and if kept longer could reach 84 cm (at least 9.1 kg). There was a considerable market in London, where a carp between 33 cm and 40 cm could fetch 12d (North 1714: 63–64, 90).

Estate records of the period confirm that many park and garden ponds were still expected to contribute fish for the table. Ponds within parks continued to fulfil a dual function, providing fish in addition to water for deer. John Norden's survey of Windsor Great Park in 1607 shows half-a-dozen dammed ponds on streams within the park (Roberts 1997: 246–247). Lord Wharton's journal records yields from several of his park fishponds at Upper Winchendon in 1686 and 1700: several hundred carp and tench are mentioned, along with 50 pike in one pond, also numerous perch and a few eels. Some of the fish were sold, others kept for restocking (Croft and Pike 1988: 264–5). The diary of Henry Hyde, 2nd Earl of Clarendon, records the drawing of seven ponds in Cornbury and Wychwood between April and September 1690, during which six large carp were harvested from one pond; the ponds were restocked with over 1,200 carp (Bond and Chambers 1988: 366). In the 1690s the Earl of Bedford regularly stocked his ponds at Woburn with live pike and perch carried by barge from Thorney in the Fens (Thomson 1940: 159). The continuing utility of Dyrham Park's ornamental ponds is demonstrated by an account of August 1710, which documents the numbers of carp, trout, tench and perch in each pond and estimates their yield for the table up to ten years ahead (Dennison and Iles 1985: 41, 48). The Rectory ponds at Souldern produced 31 brace of carp in 1723 (Bond and Chambers 1988: 366). Ponds on the former Wriothesley estates around Titchfield still contained carp in the 1740s (Currie 1991: 104). A chain of five ornamental fishponds in Halswell Park were stocked with goldfish, gudgeon and trout in the 1750s. Even on the grand but short-lived formal canal at Blenheim Park, Boydell's illustrations published in 1752 show figures fishing with rod and line and using a net (Bond and Tiller 1997: 81).

Excavation of a pond at Castle Bromwich revealed a brick-lined exit sluice dating from about 1730 which would have permitted the pond to be emptied; the entrance was protected by an iron grill to prevent weed and leaves from blocking the system; this pond was not on a major vista and seems to have been designed at least partly for fish (Currie 1990: 41).

Introductions of further fish species from abroad were limited between 1660 and 1800. The Crucian carp (*Carrasius carassius*), represented by one bone from Roman levels in London, has been claimed as a native species (Newdick 1979; Wheeler 2000); but it makes no further appearance before the early eighteenth century, when imports came in from Hamburg (Houghton 1879; 1984 edition: 52–4). Jethro Tull acquired further specimens from Germany to experiment with castration to improve the quality of the flesh (Couch 1865: iv, 28). However, in English conditions it remained relatively small and slow-growing. Although it was installed in some fish-ponds in south-east England (Pennant 1812) and occasionally appeared in the Thames and the Fenland drains, it remained relatively rare, probably being kept more as an ornamental curiosity than for its food value (Yarrell 1841, i: 355). Its close relative the Prussian carp (*Carassius carassius gibelio*) also appeared in fishponds in the south-east around the same time, and became better-established in the Norfolk Broads and elsewhere (Pennant 1812; Houghton 1879; 1984 edition: 54). The goldfish (*Carassius auratus*), which has a natural range extending from China and Japan into eastern Europe, may have been introduced into western Europe as early as 1611, but remained unknown in Britain until the early eighteenth century. In 1730 Sir Matthew Decker, a director of the East India Company and Lord Mayor of London, having acquired many specimens from the continent, presented them to various friends elsewhere in England. By 1750 goldfish were widely kept for ornament in shallow ponds (Lever 1979: 447–55).

The eighteenth and nineteenth centuries

By the 1740s the geometrically-shaped ponds associated with formal gardens were passing out of fashion. Some were abandoned, others altered, as revolutionary ideas of 'landscape' gardening encouraged the creation of larger lakes of more 'natural' appearance. Some of the new lakes, including that at Stourhead made in 1744, and Lancelot Brown's lakes at Longleat and Blenheim Park made in the 1750s and 1760s, submerged the sites of documented medieval fishponds. Landowners valuing picturesque views would have abhorred the unsightly traditional practices of draining-down, netting and leaving beds dry at regular intervals, and eighteenth-century lakes were hardly ever provided with the diversion channels necessary for those purposes. The inability to divert incoming water did lead to longer-term management problems, making it more difficult to maintain the dam and pond bed and to clear the accumulating silt. Neverthless, such lakes continued to be stocked with fish for angling, now well established as a pleasant gentlemanly pastime (Currie 1990: 42) and, increasingly, as a competitive sport. The fishing of Stonehead Lake in 1793 produced 2,000 carp 'of large dimensions', including one 8 kg specimen, while a brace of carp caught in Gratton Park weighed nearly 16 kg together (Currie 1991: 106). Two ponds within the Hafod Uchtryd demesne in Cardiganshire, documented as producing fish in the late 1780s, were also associated with ornamental components (Kerkham and Briggs 1991: 169).

The triumph of aesthetic considerations over efficient fish production was aided by eighteenth-century improvements in long-distance carriage and transport, which opened up alternative sources of supply. Daniel Defoe describes how, by the 1720s, live tench and pike from the Fenland meres were regularly transported to London by waggon in great water-butts, equipped with flaps to provide air; each night on the journey the water was changed (Defoe 1724–6; 1971 edition: 417). By the end of the century Scottish salmon were being packed in ice in boxes for transport by sea to London (Wilson 1976: 48).

A particular feature of aristocratic parkland fishing was the renewed provision of ornamental fishing pavilions in a variety of styles. In Windsor Great Park the Duke of Cumberland had a new lake created on the site of the present Virginia Water in the 1750s, and a design was produced for a Fishing Temple on an arcaded base. Before this could be built, the dam collapsed during a night of torrential rain on 1 September 1768. The lake was restored and enlarged during the 1780s. George IV enjoyed fishing here almost daily, and in 1825 engaged Jeffry Wyattville to design a new Fishing Temple in Chinese style, built on the site of the medieval lodge demolished in the early 1790s. In 1827 Frederick Crace elaborated the building, with an enriched roof design. Proposals to restore it in 1860 were rejected by Queen Victoria, and by 1870 it had been demolished and replaced with a simpler pavilion designed by Anthony Salvin to resemble a Swiss chalet. This was in turn demolished in 1936 when Edward VIII stayed at Fort Belvedere nearby, since it was thought to spoil the view of the lake (Roberts 1997: 414–422).

Many once-productive fishponds had now been abandoned. The dam of the Foss Pool in York had been removed by the mid-nineteenth century and its bed was infilled with rubbish (RCHME York 1972: 61, 138). Yet some advocates of agricultural improvement were still promoting fishponds as a contribution to the farming economy into the early nineteenth century. Arthur Young (1813: 34) praised the fishponds on Sir Christopher Willoughby's farm at Marsh Baldon, which afforded him "carp of three to six pounds, tench of one pound and perch from a half to two pounds... to be had whenever he wants them". A group of linear ponds at Merton (Oxfordshire) had been extended twice since 1763, possibly continuing in use into the early twentieth century (Chambers and Gray 1988: 118–9). A few examples were still managed by traditional means. Frensham Great Pond was still emptied every five years for fishing-out as late as 1858 (Roberts 1986: 133, 135).

Izaak Walton had praised the culinary values of a wide range of freshwater fish, but by the nineteenth century even the late medieval favourites, bream, pike and carp, had largely lost their repute. This may be related to the reduced numbers of dedicated fishponds, since it was noted that some fish tasted much better from well-managed ponds than when caught in muddy river waters (Houghton 1879; 1984 edition: 50, 83, 87). Inevitably personal preferences varied. Mrs Beeton's famous cookery book of 1869 favoured dace, gudgeon and tench, but she was unenthusiastic about pike ('flesh is generally dry') and chub ('not much esteemed'). Houghton believed gudgeon, though small, to be one of the best freshwater fish for flavour available in Britain, while loach was 'by some accounted an excellent food', but he felt unable to recommend carp, bream, chub, bleak or barbel. He also regarded Iolo Goch's *gwyniad* as insipid, fit only for salting for consumption by the poor (Houghton 1879; 1984 edition: 47–51, 60–65, 70–73, 86–89, 95–8, 100–101, 200–202, 207). Tench, chub and bleak, though once consumed, are now regarded as barely palatable.

The only freshwater fish to show a marked gain in popularity was the trout, rarely kept in fishponds before the sixteenth century. Dubravius (1599: 37–8) had described trout-ponds as a recent innovation, intended to provide fish for banquets rather than to make profit; he stated that trout ponds required cold, running water, and needed to be deeper than the usual store-ponds, with a bottom of sand and gravel. Trout was 'esteemed... for its delicacy' by Mrs Beeton, and fly fishing for trout developed as a socially prestigious sport in the chalk streams of Hampshire and other southern counties during the nineteenth century.

In the more recent development of fishponds, meeting the requirements of anglers has generally continued to take precedence over efficient supplies of fish for direct consumption. The principles of artificial breeding, discovered in France around 1420, were not applied on a significant scale until the later nineteenth century. Improvements in the production of *Salmonidae* thereafter were considerable, between 50 and 80 per cent of the hatched fish reaching the yearling stage, after which losses were small (Hall 1949).

Commercial hatcheries experimented with further species from continental Europe and North America. Frank Buckland, a founder-member of the Acclimatisation Society, was a key figure in distributing and monitoring the progress of introduced species such as the golden tench (1867) and North American brook trout or brook charr (*Salvelinus fontinalis*) (1869). Buckland was in regular contact with influential proprietors such as the Duke of Bedford, who maintained ponds at Woburn Abbey. Introductions at Woburn in the 1870s and early 1880s included the orfe or ide (*Leuciscus idus*) from central Europe, the wels or European catfish (*Silurus glanis*) from the Danube and the zander or pike-perch (*Sander lucioperca)* from Schleswig-Holstein.

Between 1884 and 1905 shipments of eggs of the North American rainbow trout (*Salmo gairdneri*), arriving at the National Fish Culture Association at Delaford Park, were distributed around various hatcheries in Britain. This tolerates a wider water temperature range and higher pollution, consumes a wider range of food and grows faster than the native brown trout, and breeding stock became established in a number of lakes and rivers. However, although conditions for successful reproduction were often achieved in managed ponds, complete naturalisation was prevented by competition and periodic viral diseases (Lever 1979: 413–38, 460–71, 488–96).

References

Aberg, F. A. (ed.) 1978. *Medieval Moated Sites* (Council for British Archaeology, Research report 17). London.

Allcroft, A. H. 1908. *Earthwork of England*. London.

Andrews, M. 1993. Theobalds palace: the gardens and park. *Garden history* 21.i: 129–149.

Anonymous 1496. *The treatyse of fysshinge with an angle*, included in the 2nd edition of *The boke of St Albans*. London.

Aston, M. 1970–72. Earthworks at the bishop's palace, Alvechurch, Worcestershire. *Transactions of Worcestershire archaeological society*, 3rd series, 3: 55–59.

Aston, M. 1982. Aspects of fishpond construction and maintenance in the 16th and 17th centuries, with particular reference to Worcestershire, pp. 257–80 in T. R. Slater and P. J. Jarvis (eds.), *Field and Forest: An Historical Geography of Warwickshire and Worcestershire*. Norwich.

Aston, M. (ed.) 1988. *Medieval Fish, Fisheries and Fishponds in England*. Oxford. (British Archaeological Reports, British series 182, 2 parts).

Aston, M. and Bond, J. 1988. Warwickshire fishponds, pp. 417–434 in M. Aston (ed.) *Medieval Fish, Fisheries and Fishponds in England*. Oxford.

Balon, E. K. 1974. *The Domestication of the Carp*. Toronto.

Barker, T. 1653. *The Art of Angling*. 2nd edition, 1657, entitled *Barker's Delight*. London.

Beeton, I. M. 1869 *The Book of Household Management*. London.

Beresford, M. W. and Hurst, J. G. 1990. *Wharram Percy Deserted Medieval Village*. London.

Beresford, M. W. and St Joseph, J. K. S. 1958. *Medieval England: An Aerial Survey*. Cambridge. 2nd edition, 1979.

Bettey, J. 1993. *Estates and the English Countryside*. London.

Blagrave, J. 1685. *The Epitome of the Art of Husbandry*... [including] *directions for use of the angle*. 4th edition, London.

Bond, J. and Chambers, R. A. 1988. Oxfordshire fishponds, pp. 353–370 in M. Aston (ed.) *Medieval Fish, Fisheries and Fishponds in England*. Oxford.

Bond, J. and Tiller, K. (eds.) 1997. *Blenheim: Landscape for a Palace*. 2nd edition, Stroud.

Bradley, R. 1721. *A General Treatise of Husbandry and Gardening*. London.

Braekman, W. L. 1980. *The Treatise on Angling in the Boke of St Albans, 1496*. Brussels.

Brown, A. E. and Taylor, C. C. 1973. The gardens at Lyveden, Northamptonshire. *Archaeological journal* 129: 154–160.

Brown, R. A., Colvin, H. M. and Taylor, A. J. 1963. *The History of the King's Works* volumes 1 and 2, *The Middle Ages*. London.

Chambers, R. A. 1975. Three fishponds at Thame, Oxfordshire, 1973. *Oxoniensia* 40: 238–246.

Chambers, R. A. 1977. Observations at Somerton, 1973. *Oxoniensia* 42: 216–225.

Chambers, R. A. and Gray, M. 1988. The excavation of fishponds, pp. 113–135 in M. Aston (ed.) *Medieval Fish, Fisheries and Fishponds in England*. Oxford.

Close R. = *Calendar of Close Rolls*. London.

Couch, J. 1862–65. *A History of Fishes of the British Isles*, 4 volumes. London.

Creighton, O. H. 2002. *Castles and Landscapes: Power, Community and Fortification in Medieval England*. London.

Croft, R. A. and Pike, A. R. 1988. Buckinghamshire fishponds and river fisheries, pp. 229–266 in M. Aston (ed.) *Medieval Fish, Fisheries and Fishponds in England*. Oxford.

Currie, C. K. 1988. Medieval fishponds in Hampshire, pp. 267–289 in M. Aston (ed.) *Medieval Fish, Fisheries and Fishponds in England*. Oxford.

Currie, C. K. 1990. Fishponds as garden features, c.1550–1750. *Garden History* 18: 22–46.

Currie, C. K. 1991. The early history of the carp and its economic significance in England. *Agricultural History Review* 39: 97–107.

DB = *Domesday Book*: among modern English translations, the most convenient is Williams, A. and Martin, G. H. (eds.), 1992 *Domesday Book: a complete translation* London.

Davis, D. 1966. *A History of Shopping*. London.

Defoe, D. 1724–26. *A Tour Through the Whole Island of Great Britain*. 1971 abridged edition, Harmondsworth.

Dennison, E. and Iles, R. 1985. Medieval fishponds in Avon. *Bristol and Avon Archaeology* 4: 34–51.

Dennys, J. 1613. *The Secrets of Angling*. London.

Dubravius, J. 1599. *A new booke of good Husbandry... conteining the Order and maner of making Fish-pondes....* London.

Dyer, C. C. 1988. The consumption of fresh-water fish in medieval England, pp. 27–38 in Aston, M. (ed.) *Medieval fish, fisheries and fishponds in England*, Oxford.

Everson, P. 1989. The gardens of Campden House, Chipping Campden, Gloucestershire. *Garden History* 17.ii: 109–121.

Everson, P. 1991. Field survey and garden earthworks, pp. 6–19 in Brown, A.E. (ed.) *Garden Archaeology*. London (Council for British Archaeology, Research report 78).

Everson, P. L., Taylor, C. C. and Dunn, C. J. 1991. *Change and Continuity: Rural Settlement in North-West Lincolnshire*. London.

Furnivall, F. J. (ed.) 1870. *A Compendyous Regyment or A dyetary of helth made in Mountpyllier compyled by Andrewe Boorde of physycke doctoure*. London.

Gilbert, W. 1676. *The Angler's Delight, containing the whole art of neat and clean angling*. London.

Glasse, H. 1747. *The Art of Cookery*. 2nd edition, London.

Hague, D. B. and Warhurst, C. 1966. Excavations at Sycharth Castle, Denbighshire, 1962–3. *Archaeologia Cambrensis* 115: 108–127.

Hall, C. A. 1949. *Ponds and Fish Culture*. London.

Hale, J. 1756. *The Complete Body of Husbandry*. London.

Hartley, R. F. 1983. *The Medieval Earthworks of Rutland: a survey*. Leicester.

Harvey, J. 1975. *Mediaeval Craftsmen*. London.

Harvey, P. D. A. 1965. *A Medieval Oxfordshire Village: Cuxham, 1240 to 1400*. London.

Henderson, P. 1992. Sir Francis Bacon's water gardens at Gorhambury. *Garden History* 20: 116–131.

Hickling, C. F. 1962. *Fish Culture*. London.

Hoffmann, R. C. 1985. Fishing for sport in medieval Europe: new evidence. *Speculum* 60: 877–902.

Hoffmann, R. C. 1994. Medieval Cistercian fisheries, natural and artificial, pp. 401–414 in L. Pressouyre (ed.) *L'espace Cistercien*. Paris.

Hollings, M. (ed.) 1934, 1937, 1939, 1940. *The Red Book of Worcester*. 4 volumes, Worcestershire Historical Society.

Holt, R. 1988. *The Mills of Medieval England*. Oxford.

Houghton, W. 1879. *British Fresh-Water Fishes*. reprinted 1984, London.

Hudson, T. (ed.) 1841. *Manners and Household expenses of England in the 13th and 15th Centuries*. London.

James, J. 1712. *The Theory and Practice of Gardening*. London.

Kerkham, C. and Briggs, S. 1991 A review of the archaeological potential of the Hafod landscape, Cardiganshire, pp. 160–174 in A. E. Brown (ed.) *Garden Archaeology*. London.

L. & P. Henry VIII: *Letters and papers of Henry VIII*. London.

Lawrance, P. 1982. Animal bones, pp. 275–296 in J. G. Coad and A. D. F. Streeten, Excavations at Castle Acre Castle, Norfolk, 1972–77. *Archaeological journal* 139: 138–301.

Lawrance, P. 1987. Animal bones, pp. 297–302 in J. G. Coad, A. D. F. Streeten and R. Warmington, Excavations at Castle Acre, Norfolk, 1975–1982. *Archaeological journal* 144: 256–307.

Lawson, W. 1617. *A new orchard and garden*. London.

Lever, C. 1979. *The naturalised animals of the British Isles*. London.

Liberate R. = *Calendar of Liberate Rolls*. London.

Liddiard, R. 2005. *Castles in Context: Power, Symbolism and the Landscape, 1066 to 1500*. Macclesfield.

McDonald, J. 1963. *The Origins of Angling*. New York.

McDonnell, J. 1981. *Inland fisheries in medieval Yorkshire, 1066–1300*. York (Borthwick Papers 60, Borthwick Institute of Historical Research, University of York).

Markham, G. 1608. *The Whole Art of Husbandry*, Book IV: *of poultrie, fowle, fishe and bees*, [including] *the art of angling…* London.

Markham, G. 1611. *Country Contentments; or, the husbandman's recreations…* [including] *the whole art of angling…* 5th edition, 1633. London.

Markham, G. 1613. *The English Husbandman…* [including] *discourse of the general art of fishing with an angle*. London.

Markham, G. 1614. *Cheap and Good Husbandry…* [including] *the making of fish-ponds and the taking of all sorts of fish…* 3rd edition, 1623. London.

Markham, G. 1631. *A way to get wealth*, 5th edition. London.

Markham, G. 1635. *The Pleasures of Princes, or good men's recreations, containing a discourse of the generall art of fishing with the angle or otherwise…* London.

Mascall, L. 1590. *A Booke of Fishing with Hooke and Line*. London.

Moorhouse, S. 1981. Fishponds and fisheries, pp. 743–51 in M. L. Faull and S. Moorhouse (eds.) *West Yorkshire: an archaeological survey to AD 1500*, volume 3, *The rural medieval landscape*. Wakefield.

Moorhouse, S. 1988. Medieval fishponds: some thoughts, pp. 475–484 in M. Aston (ed.) *Medieval Fish, Fisheries and Fishponds in England*. Oxford.

Mortimer, J. 1707. *The Whole Art of Husbandry*. London.

Newdick, J. 1979. *The complete freshwater fishes of the British Isles*. London.

North, R. 1714. *A discourse of fish and fish-ponds*. London.

Patent R. = *Calendar of Patent Rolls*. London.

Pennant, T. 1812. *British Zoology*, vol. 4. London.

RCHME Northants. 1975. Royal Commission on Historical Monuments, England *An inventory of the historical monuments in the county of Northampton*, vol. 1: *Archaeological sites in north-east Northamptonshire*. London.

RCHME Northants. 1979. Royal Commission on Historical Monuments, England *An inventory of the historical monuments in the county of Northampton*, vol. 2: *Archaeological sites in central Northamptonshire*. London.

RCHME York 1972. Royal Commission on Historical Monuments, England, *An inventory of the historical monuments in the city of York*, 2: *The defences*. London.

Richardson, H. G. and Sayles, G. O. (eds.) 1955. *Fleta*, vol. 2. London (Selden Society, volume 72).

Roberts, B. K. 1966. Medieval fishponds. *Amateur historian* 7.iv: 119–26.

Roberts, B. K. 1988. The re-discovery of fishponds, pp. 9–16 in M. Aston (ed.) *Medieval Fish, Fisheries and Fishponds in England*. Oxford.

Roberts, E. 1986. The Bishop of Winchester's fishponds in Hampshire, 1150–1400: their development, function and management. *Proceedings of Hampshire Field Club* 42: 125–138.

Roberts, J. 1997. *Royal landscape: the gardens and parks of Windsor*. New Haven.

Robinson, F. W. (ed.) 1950. *The Works of Geoffrey Chaucer*. 2nd edition, Oxford.

SOED, 1983. *The Shorter Oxford English Dictionary*. London.

Steane, J. M. 1970. The medieval fishponds of Northamptonshire. *Northamptonshire Past and Present*, 4: 299–310.

Steane, J. M. 1988. The royal fishponds of medieval England, pp. 39–68 in M. Aston (ed.) *Medieval Fish, Fisheries and Fishponds in England*. Oxford.

Steane, J. M. and Foreman, M. 1988. Medieval fishing tackle, pp. 137–186 in M. Aston (ed.) *Medieval Fish, Fisheries and Fishponds in England*. Oxford.

Switzer, S. 1718. *Ichnographia rustica, or, The nobleman, gentleman and gardener's recreation*. London.

Switzer, S. 1734. *A Universal System of Water and Water-Works, Philosophical and Practical*. 2 volumes, London.

Taverner, J. 1600. *Certaine experiments concerning fish and fruite*. London.

Taylor, A. 1977. A culvert beneath the sea bank at Newton near Wisbech. *Proceedings of Cambridge Antiquarian Society* 67: 63–66.

Taylor, C. C. 1989. Somersham Palace, Cambridgeshire: a medieval landscape for pleasure? pp. 211–224 in M. Bowden, D. Mackay and P. Topping (eds.) *From Cornwall to Caithness: Some Aspects of British Field Archaeology*. Oxford. (British Archaeological Reports, British series, 209).

Taylor, C. C. 2000. Medieval ornamental landscapes. *Landscapes* 1: 38–55.

Taylor, C. C., Everson, P. and Wilson-North, R. 1990. Bodiam Castle, Sussex. *Medieval archaeology* 34: 155–157.

Thirsk, J. 1967. Farming techniques, pp. 161–199 in J. Thirsk (ed.) *The Agrarian History of England and Wales*, vol. 4, *1500–1640*. Cambridge.

Thomas, R. 1981. Stone weights from the Thames. *Oxoniensia* 46: 129–133.

Thomas, G. S. 1940. *Life in a Noble Household, 1641–1700*. London.

Thompson, M. W. 1965. Two levels of the mere at Kenilworth Castle, Warwickshire. *Medieval archaeology* 9: 156–161.

Thorpe, L. (ed.) 1978. Gerald of Wales, *The Journey through Wales / The Description of Wales*. Harmondsworth.

Thurley, S. 1993. *The Royal Palaces of Tudor England*. New Haven.

Van Siclen, G. W. 1880. *An American edition of the Treatyse of fysshynge wyth an angle, from the Boke of St Albans... A.D. 1496*. New York.

Walton, I. 1653. *The compleat angler, or the contemplative man's recreation*. London. 5th edition, 1676 (retitled *The universal angler*), reprinted 1906: London.

Wilson, C. A. 1976. *Food and drink in Britain*. Harmondsworth.

Venables, R. 1662. *The experienced angler, or angling improved*. London.

Wheeler, A. 2000. The status of the crucian carp, *Carassius carassius* (L.) in the U.K. *Fisheries management and ecology* 7: 315–322.

Whittle, E. H. 1989. The renaissance gardens of Raglan Castle. *Garden history* 17: 83–94.

Whittle, E. and Taylor, C. 1994. The early seventeenth-century gardens of Tackley, Oxfordshire. *Garden History* 22: 37–63.

Worlidge, J. 1669. *Systema agricultura*. 3rd edition, 1681, London.

Yarrell, W. 1841. *A History of British Fishes*. 2nd edition, London.

Young, A. 1813. *General view of the agriculture of Oxfordshire*. London.

Zeepvat, R. 1988. Fishponds in Roman Britain', pp. 17–26 in M. Aston (ed.) *Medieval Fish, Fisheries and Fishponds in England*, Oxford.

Figure 7.1: Royal, episcopal and secular fishponds in England and Wales: location of sites mentioned in the text.

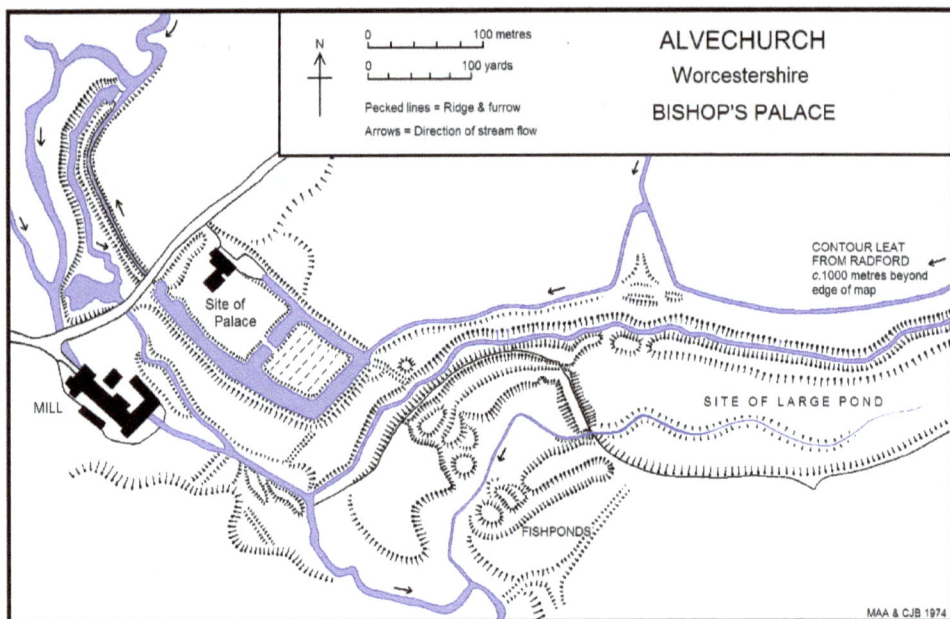

Figure 7.2: Alvechurch, Worcestershire: moats, fishponds and leats associated with a medieval palace of the Bishops of Worcester.

Figure 7.3: Harrington, Northamptonshire: prominent earthworks of a group of three medieval valley fishponds, showing the bypass leat around the central pond. (Photo: Ed Dennison)

Figure 7.4: Tackley, Oxfordshire: triangular pond, part of a complex of geometrically-shaped fishponds laidout by John Harborne shortly after 1612. (Photo Ed Denison)

Figure 7.5: Old fishpond at the Manor House, Long Clawson, Leicestershire: a fishpond of medieval origin still stocke today. (Photo: Nicholas Redman, 2012)

Contributors

James Bond, The Anchorage, Coast Road, Walton-in-Gordano, Clevedon, North Somerset, England.

Madeleine Bonow, School of Natural Science, Technology and Environmental Studies, Södertörn University, Sweden.

Stanisław Cios, Department of Global Energy, Ministry of Foreign Affairs, Warsaw, Poland.

Erik Hofmeister, The State and University Library, Aarhus, Denmark.

Anne Karin Hufthammer, Department of Natural History, The University Museum, University of Bergen.

Dagfinn Moe, Department of Natural History, The University Museum, University of Bergen.

Håkan Olsén, School of Natural Science, Technology and Environmental Studies, Södertörn University, Sweden.

Ingvar Svanberg, Uppsala Centre for Russian and Eurasian Studies, Uppsala University, Sweden.

This book falls within the framework of the research project "The Story of Crucian Carp in the Baltic Sea Region: History and a Possible Future" (2010–2016) sponsored by the The Foundation for Baltic and East European Studies (Östersjöstiftelsen) (dnr. A017-09; K. Håkan Olsén). Within this project the following studies have been completed:

Publications in English

Stefan Lundberg & Ingvar Svanberg, The stone loach in Stockholm, Sweden and royal fish-ponds in the seventeenth and eighteenth centuries. *Archives of Natural History* 37 (2010), pp. 150–160.

Ingvar Svanberg, Madeleine Bonow & Håkan Olsén, Fish ponds in Scania, and Linnaeus's attempt to promote aquaculture in Sweden. *Svenska Linnésäll-skapets Årsskrift* (2012), pp. 83–98.

Johanna Wouters, Sven Janson, Věra Lusková & K. Håkan Olsén, Molecular identification of hybrids of the invasive gibel carp *Carassius auratus gibelio* and crucian carp *Carassius carassius* in Swedish waters. *Journal of Fish Biology* 80:7 (2012), pp. 2595–2604.

Ingvar Svanberg & Stanisław Cios, Petrus Magni and the history of fresh-water aquaculture in the later Middle Ages. *Archives of Natural History* 41:1 (2014), pp. 124–30.

Sven Janson, Johanna Wouters, Madeleine Bonow, Ingvar Svanberg & Håkan Olsén, Population genetic structure of crucian carp (*Carassius carassius*) in man-made ponds and wild populations in Sweden. *Aquatic International* 23 (2015), pp. 359–368.

Håkan Olsén & Ingvar Svanberg, Introduction, pp. 11–27, in *History of Aquaculture in Northern Europe*, peds. M. Bonow, H. Olsén and I. Svanberg. Stockholm (2016).

Madeleine Bonow, Håkan Olsén and Ingvar Svanberg (eds.), Historical Aquaculture in Northern Europe. Huddinge: Södertörn University, 2016. 206 pages.

Madeleine Bonow & Ingvar Svanberg, Historical Pond-Breeding of Cyprinids in Sweden and Finland, pp. 91–123 in *Historical Aquaculture in Northern Europe*, eds. Madeleine Bonow, Håkan Olsén and Ingvar Svanberg.Huddinge: Södertörn University, 2016.

Madeleine Bonow, Stanisław Cios & Ingvar Svanberg, Fishponds in the Baltic States: Historical Cyprinid Culture in Estonia, Latvia and Lithuania, pp. 145–162 in *Historical Aquaculture in Northern Europe*, eds. Madeleine Bonow, Håkan Olsén and Ingvar Svanberg. Huddinge: Södertörn University, 2016.

Daniel L. Jeffries, Gordon H. Copp, Lori Lawson Handley, K. Håkan Olsén, Carl, D. Sayer, & Bernd Hänfling. Comparing RADseq and microsatellites to infer complex phylogeographic patterns, an empirical perspective in the crucian carp *Carassius carassius*, L. *Molecular Ecology*, (2016) doi: 10.1111/mec.13613.

K. Håkan Olsén & Torbjörn Lundh. Feeding stimulants in an omnivorous species, crucian carp *Carassius carassius* (Linnaeus 1758). *Aquaculture Reports* 4 (2016), pp. 66–73.

Publications in Swedish

Madeleine Bonow & Ingvar Svanberg, »Säj får jag dig bjuda ur sumpen en sprittande ruda«: en bortglömde läckerhet från gångna tiders prästgårdskök, pp. 147–169 in *Gastronomins (politiska) geografi*, red. Madeleine Bonow & Paulina Rytkönen (=Årsboken Ymer årg. 141). Stockholm: Svenska Sällskapet för Antropologi och Geografi, (2011).

Madeleine Bonow & Ingvar Svanberg, Uppländska ruddammar: ett bidrag till akvakulturens historia. *Uppland: Årsbok* (2012), pp. 123–152.

Madeleine Bonow & Ingvar Svanberg, »Rudor finnas öfverflödigt«: Fiskdammar vid svenska prästgårdar på 1600- och 1700-talen. *Saga och Sed: Kungl. Gustav Adolfs Akademiens Årsbok* 2013 (2014), pp. 111–131.

Gunilla Lindell & Ingvar Svanberg, *Insjöfisk: recept och kulturhistoria*. Stockholm: Molin & Sorgenfrei, (2014). 192 pages.

Madeleine Bonow & Ingvar Svanberg, Urbana fiskdammar i 1600- och 1700-talets Sverige: strödda notiser om akvakultur i stadsmiljö. *Rig* 97:4 (2015), pp. 215–222.

Madeleine Bonow & Ingvar Svanberg, Monastiska fiskdammar i det senmedel-tida Sverige, pp. 266–284 in *Biskop Brasks måltider: svensk mat mellan medeltid och renässans*, red. Magnus Gröntoft m.fl. Stockholm: Atlantis, (2016).

www.ingramcontent.com/pod-product-compliance
Lightning Source LLC
Chambersburg PA
CBHW052129270326
41930CB00012B/2816